Design of Low-Power Coarse-Grained Reconfigurable Architectures

Design of Low-Power Coarse-Grained Reconfigurable Architectures

Yoonjin Kim
Rabi N. Mahapatra

CRC Press
Taylor & Francis Group
Boca Raton   London   New York

CRC Press is an imprint of the
Taylor & Francis Group an **informa** business
A CHAPMAN & HALL BOOK

CRC Press
Taylor & Francis Group
6000 Broken Sound Parkway NW, Suite 300
Boca Raton, FL 33487-2742

First issued in paperback 2017

© 2011 by Taylor and Francis Group, LLC
CRC Press is an imprint of Taylor & Francis Group, an Informa business

No claim to original U.S. Government works

ISBN 13: 978-1-138-11352-7 (pbk)
ISBN 13: 978-1-4398-2510-5 (hbk)

Visit the Taylor & Francis Web site at
http://www.taylorandfrancis.com

and the CRC Press Web site at
http://www.crcpress.com

Contents

List of Figures

List of Tables

Preface

Application-specific optimization of embedded systems becomes inevitable to satisfy the market demand for designers to meet tighter constraints on cost, performance and power. On the other hand, the flexibility of a system is also important to accommodate the short time-to-market requirements for embedded systems. To compromise these incompatible demands, coarse-grained reconfigurable architecture (CGRA) has emerged as a suitable solution. A typical CGRA requires many processing elements (PEs) and a configuration cache for reconfiguration of its PE array. However, such a structure consumes significant area and power. Therefore, designing cost-effective CGRA has been a serious concern for reliability of CGRA-based embedded systems.

As an effort to provide such cost-effective design, the first half of this book focuses on reducing power in the configuration cache. For power saving in the configuration cache, a *low-power reconfiguration technique* is presented based on reusable context pipelining achieved by merging the concept of context reuse into context pipelining. In addition, we propose *dynamic context compression* capable of supporting only required bits of the context words set to enable and the redundant bits set to disable. Finally, we provide *dynamic context management* capable of reducing power consumption in configuration cache by controlling a read/write operation of the redundant context words.

In the second part of this book, we focus on designing a cost-effective PE array to reduce area and power. For area and power saving in a PE array, we devise a *cost-effective array fabric* that addresses novel rearrangement of processing elements and their interconnection designs to reduce area and power consumption. In addition, *hierarchical reconfigurable computing arrays* are proposed consisting of two reconfigurable computing blocks with two types of communication structure together. The two computing blocks have shared critical resources and such a sharing structure provides efficient communication interface between them with reducing overall area. Based on the proposed design approaches, a CGRA combining the multiple design schemes is shown to verify the synergy effect of the integrated approach.

Audience for This Book

This book is intended for computer professionals, graduate students, and advanced undergraduates who need to understand issues involved in designing and constructing embedded systems. The reader is assumed to have had introductory courses in digital system, VLSI design, computer architecture, or equivalent work experience.

Chapter 1

Introduction

1.1 Coarse-Grained Reconfigurable Architecture

With the growing demand for high quality multimedia, especially over portable media, there has been continuous development on more sophisticated algorithms for audio, video, and graphics processing. These algorithms have the characteristics of data-intensive computation of high complexity. For such applications, we can consider two extreme approaches to implementation: software running on a general purpose processor and hardware in the form of Application-Specific Integrated Circuit (ASIC). In the case of general purpose processor, it is flexible enough to support various applications but may not provide sufficient performance to cope with the complexity of the applications. In the case of ASIC, we can optimize best in terms of power and performance but only for a specific application. With a coarse-grained reconfigurable architecture (CGRA), we can take advantage of the above two approaches. This architecture has higher performance level than general purpose processor and wider applicability than ASIC.

As the market pressure of embedded systems compels the designer to meet tighter constraints on cost, performance, and power, the application specific optimization of a system becomes inevitable. On the other hand, the flexibility of a system is also important to accommodate rapidly changing consumer needs. To compromise these incompatible demands, domain-specific design is focused on as a suitable solution for recent embedded systems. Coarse-grained reconfigurable architecture is the very domain-specific design in that it can boost the performance by adopting specific hardware engines while it can be reconfigured to adapt to ever-changing characteristics of the applications.

Typically, a CGRA consists of a main processor, a Reconfigurable Array Architecture (RAA), and their interface as Figure 1.1. The RAA has identical processing elements (PEs) containing functional units and a few storage units such as ALU, multiplier, shifter and register file. The data buffer provides operand data to PE array through a high-bandwidth data bus. The configuration cache (or context memory) stores the context words used for configuring the PE array elements. The context register between a PE and a cache element (CE) in configuration cache is used to keep the cache access path from being the critical path of the CGRA.

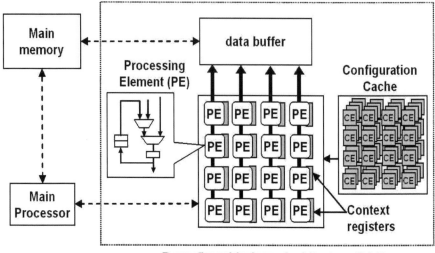

Reconfigurable Array Architecture (RAA)

FIGURE 1.1: Block diagram of general CGRA.

Unlike FPGA (most typical of a fine-grained reconfigurable architecture), which are built with bit-level configurable logic blocks (CLBs), CGRA is built with PEs, which are word-level configurable functional blocks. By raising the granularity of operations from a bit to a word, CGRA can improve on the speed and the performance as well as the resource utilization for compute-intensive applications. Another consequence of this raised granularity is that whereas FPGA can be used for implementing any digital circuits, CGRA is targeted only for a limited set of applications, although different CGRAs may target different application domains. Still, CGRA retains the idea of "reprogrammable hardware" in the reprogrammable interconnects as well as in the configurable functional blocks (i.e., PEs). Moreover, since the amount of the configuration bit-stream is greatly reduced through the raised granularity, the configuration can be actually changed even at the runtime very fast. Most of the CGRAs feature single-cycle configuration change, fetching the configuration data from a distributed local cache. This unique combination of efficiency and flexibility, which is the main advantage of CGRA, explains an evaluation result [9] that under certain conditions CGRAs are actually more cost-effective for wireless communication applications than alternatives such as FPGA implementations as well as DSP architectures. It is worth mentioning that the improved efficiency of CGRAs in terms of the performance and flexibility is a result of the architecture specialization for compute-intensive applications.

In spite of the above advantages, the deployment of CGRA is prohibitive due to its significant area and power consumption. This is due to the fact that CGRA is composed of several memory components and the array of many processing elements including ALU, multiplier and divider, etc. Especially, processing element (PE) array occupies most of the area and consumes most of the power in the system to support flexibility and high performance. Therefore, reducing area and power consumption in the PE array has been a serious concern for the adoption of CGRA.

1.2 Objective and Approach

This book explores the problem of reducing area and power in CGRA based on architecture optimization. To provide cost-effective CGRA design, the following questions are considered.

- How to reduce area and power consumption in CGRA? For power saving in CGRA, we should obtain area and power breakdown data of CGRA to identify area and power-dominant components. Then the components may be optimized for area and power by removing redundancies of CGRA wasting area and power. Such redundancies may depend on the characteristics of computation model or applications.

- How to design cost-effective CGRA with non-sacrificing or enhancing performance? Ultimately, the goals of designing cost-effective CGRA is that proposed approaches do not cause performance degradation with saving area and power. It means that the proposed cost-effective CGRA keeps original functionality of CGRA intact and does not increase critical path delay. In addition, the performance may be enhanced by optimizing the performance bottleneck with keeping the area and power-efficient approaches.

In this book, these central questions are addressed for area/power-critical components of CGRA and we suggest new frameworks to achieve these goals. The validation of the proposed approaches is demonstrated through the use of real application benchmarks and gate level simulations.

1.3 Overview of the Book's Contents

This volume is divided into 11 chapters as follows:

- **Chapter 1. Introduction**
  This chapter introduces general characteristics of Coarse-Grained Reconfigurable Architecture (CGRA). In addition we present the contribution and the organization of this book.

- **Chapter 2. Trends in CGRA**
  Chapter 2 describes CGRA background and related works of this book.

- **Chapter 3. CGRA for High Performance and Flexibility**
  We describe the characteristics of typical CGRAs that achieve both high performance and flexibility. We compare the CGRAs with other computing cores such as ASIC, FPGA and general purpose processor.

- **Chapter 4. BASE CGRA Implementation**
  We have first designed a conventional CGRA as the base architecture and implemented it at the RT-level. This conventional architecture will be used throughout this book as a reference for quantitative comparison with our cost-effective approaches.

- **Chapter 5. Power Consumption in CGRA**
  We describe power consumption in CGRA and why power consumption has been a serious concern for reliability of CGRA-based embedded systems.

- **Chapter 6. Low-Power Reconfiguration Technique**
  It presents a novel power-conscious architectural technique called *reusable context pipelining* (RCP) for CGRA to close the power-performance gap between low-power-oriented spatial mapping and high performance-oriented temporal mapping prevailing in existing CGRA architectures. A new configuration cache structure has been proposed to support reusable context pipelining with negligible overheads. The temporal mapping with RCP has been shown to be a universal approach in reducing power and enhancing performance for CGRA.

- **Chapter 7. Dynamic Context Compression for Low-Power CGRA**
  A new design flow for CGRA design has been proposed to generate architecture specifications that are required for modifying configuration cache dynamically. Design methodology for dynamically compressible context architecture and a new cache structure to support the configurability are being presented to reduce the power consumption in configuration cache without performance degradation.

- **Chapter 8. Dynamic Context Management for Low-Power CGRA**
  It presents a novel control mechanism of configuration cache called dynamic context management to reduce the power consumption in configuration cache without performance degradation. A new configuration cache structure is proposed to support dynamic context management.

- **Chapter 9. Cost-Effective Array Fabric**
  A novel array fabric design exploration method has been proposed to

generate cost-effective reconfigurable array structure. Novel rearrangement of processing elements and their interconnection designs are introduced for CGRA to reduce area and power consumption without any performance degradation.

- **Chapter 10. Hierarchical Reconfigurable Computing Arrays**
  A new reconfigurable computing hierarchy has been proposed to design cost-effective CGRA-based embedded systems. Efficient communication structure between processor and reconfigurable computing blocks is introduced to reduce performance bottleneck in the CGRA-based architecture.

- **Chapter 11. Integrated Approaches**
  Chapter 11 presents an integrated approach to merge the multiple design schemes presented in the previous chapters. A case study is shown to verify the synergy effect of combining the multiple design schemes.

Chapter 2

Trends in CGRA

2.1 Introduction

A recent trend in the architectural platforms for embedded systems is the adoption of reconfigurable computing elements for cost, performance, and flexibility issues [30]. Coarse-Grained Reconfigurable Architectures (CGRAs) [30] exploit both the flexibility and efficiency, and are shown to be a generally better solution for compute-intensive applications than fine-grained reconfigurable architectures. Many kinds of coarse-grained reconfigurable architecture have been proposed with the increasing interests in reconfigurable computing [30] but such architectures are based on 2D array of ALU-like datapath blocks. These are particularly interesting due to the wide acceptance in recent reconfigurable processors as well as their expected high performance for many heavy-load applications in the domains of signal processing, multimedia, communication, security, and so on. In this chapter, we introduce the CGRA research trends in recent years in four aspects: architecture, design space exploration, code compilation & mapping and physical implementation.

2.2 Architecture

In [30], Hartenstein summarized many CGRAs that had been suggested until 2001. Since then, many more new CGRAs have been continuously proposed and evolved [16, 17, 19, 34, 36, 37, 41–43, 54, 56, 68, 74]. Most of them comprise a fixed set of specialized processing elements (PEs) and interconnection fabrics between them. The run-time control of the operation of each PE and the interconnection provides the reconfigurability. In this section, we describe entire structure, PE array fabric and memory configuration of some representative CGRA examples.

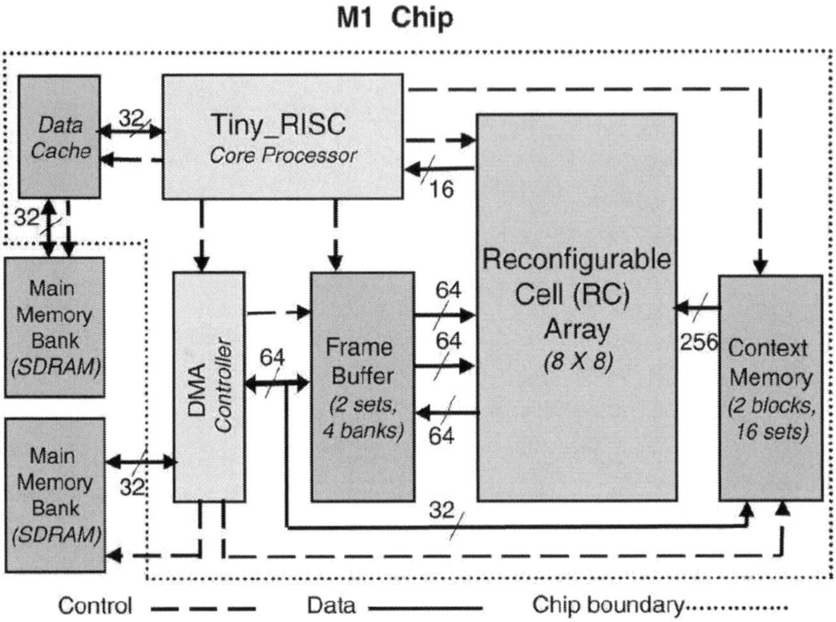

M1 Chip

FIGURE 2.1: Components of MorphoSys implementation (M1 chip). (From H. Singh, "MorphoSys: An integrated reconfigurable system for data-parallel and computation-intensive applications. *IEEE Transactions on Computers*, © 2000 IEEE. With permission.)

2.2.1   MorphoSys

The MorphoSys [75] architecture comprises five major components: the Reconfigurable Cell Array (RC Array), control processor (TinyRISC), Context Memory, Frame Buffer and a DMA Controller. Figure 2.1 shows the organization of the integrated MorphoSys reconfigurable computing system.

2.2.1.1   RC Array

The main component of MorphoSys is the 8x8 RC Array, shown in Figure 2.2. This configuration is chosen to maximally utilize the parallelism inherent in an application, which in turn enhances throughput. The RC Array follows the SIMD model of computation. All RCs in the same row/column share same configuration data (context). However, each RC operates on different data. Sharing the context across a row/column is useful for data-parallel applications.

The RC Array interconnection network consists of three hierarchical levels. Figure 2.2 shows the nearest neighbor layer that connects the RCs in a two-dimensional mesh. Thus, each RC can access data from any of its row/column neighbors (North, South, West, East neighbors). Figure 2.2 also depicts a

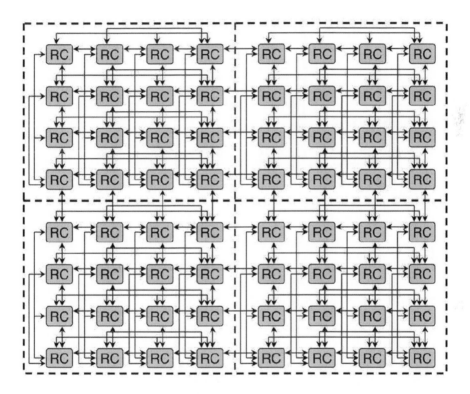

FIGURE 2.2: MorphoSys 8x8 RC array interconnection structure. (From H. Singh, "MorphoSys: An integrated reconfigurable system for data-parallel and computation-intensive applications. *IEEE Transactions on Computers*, © 2000 IEEE. With permission.)

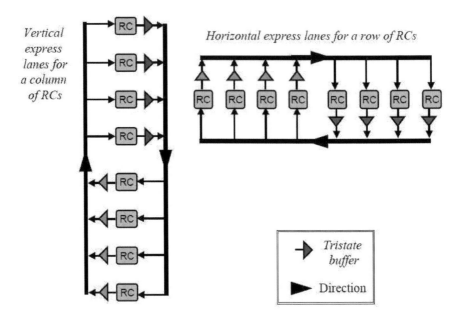

FIGURE 2.3: Express lane connectivity. (From H. Singh, "MorphoSys: An integrated reconfigurable system for data-parallel and computation-intensive applications. *IEEE Transactions on Computers*, © 2000 IEEE. With permission.)

second layer, which provides complete row and column connectivity within a quadrant. In the current MorphoSys specification, the RC array has four quadrants (from Quad 0 to Quad 3) as shown in Figure 2.2. Therefore, each RC can access the output from any other RC in its row/column in the same quadrant. At the higher or third layer, there are connections between adjacent quadrants. These buses, also called express lanes, run across rows as well as columns. The express lanes support inter-quadrant connectivity. The complete connectivity of the express lanes is depicted in Figure 2.3. Therefore, any cell in a quadrant can access one RC output in the same row/column of an adjacent quadrant at a time. The express lanes greatly enhance global connectivity. Some irregular communication patterns, that otherwise would require extensive interconnections, can be handled quite efficiently. The programmability of the interconnection network is achieved by controlling (via the context word) the input multiplexers MUX A and MUX B in each RC.

Each RC incorporates an ALU-multiplier, a shift unit, input muxes and a register file as shown in Figure 2.4. The multiplier is included since many target applications require integer multiplication. In addition, there is a context register that is used to store the current context word and provide con-

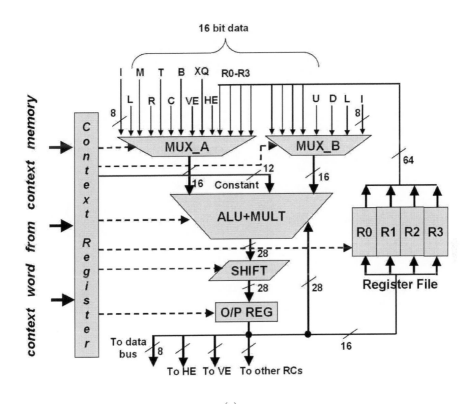

(a)

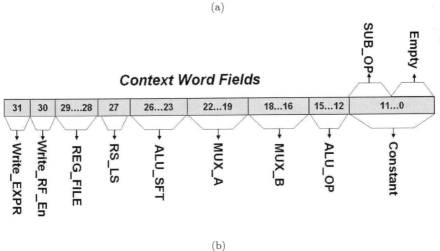

(b)

FIGURE 2.4: Reconfigurable cell (RC) structure and context architecture of MorphoSys: (a) RC structure; (b) context architecture. (From H. Singh, "MorphoSys: An integrated reconfigurable system for data-parallel and computation-intensive applications. *IEEE Transactions on Computers*, © 2000 IEEE. With permission.)

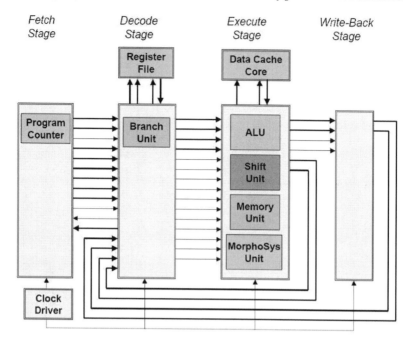

FIGURE 2.5: TinyRISC pipeline stage. (From H. Singh, "Reconfigurable Architectures for Multimedia and Data-Parallel Application Domains," Dissertation in University of California, Irvine, University of California, Irvine, 2000. With permission.)

trol/configuration signals to the RC components (namely the ALU multiplier, shift unit and the input multiplexers).

2.2.1.2   TinyRISC Processor

Since most target applications involve some sequential processing, a RISC processor, TinyRISC, is included in the system. This is a MIPS-like processor with a 4-stage scalar pipeline as shown in Figure 2.5. It has a 32-bit ALU, register file and an on-chip data cache memory. This processor also coordinates system operation and controls its interface with the external world. This is made possible by addition of specific instructions (besides the standard RISC instructions) to the TinyRISC ISA. These instructions initiate data transfers between main memory and MorphoSys components, and control execution of the RC array.

2.2.1.3   Frame Buffer and DMA

The high parallelism of the RC array would be ineffective if the memory interface is unable to transfer data at an adequate rate. Therefore, a high-speed memory interface consisting of a streaming buffer (frame buffer) and a

DMA controller is incorporated in the system. The frame buffer has two sets, which work in complementary fashion to enable overlap of data transfers with RC Array execution.

2.2.1.4 Context Memory

The Context Memory (or CM for simplicity) stores the configuration program (context) for the RC Array. As depicted in Figure 2.6, the CM is logically organized into two context blocks, called column context block and row context block. By its turn, each context block contains eight context sets. Finally, each context set has 16 context words. Context is broadcast on a row or column. Context words from one row context block are broadcast along the rows, while context words from the column context block are broadcast along the columns. In each context block, a context set is associated with a specific row or column of the RC Array. The context word from a context set is broadcast to all eight RCs in the corresponding row (column). Thus, all RCs in a row (column) share a context word and perform the same operation.

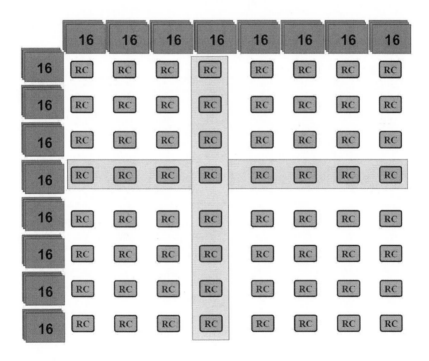

FIGURE 2.6: Organization of context memory. (From H. Singh, "Reconfigurable Architectures for Multimedia and Data-Parallel Application Domains," Dissertation in University of California, Irvine, University of California, Irvine, 2000. With permission.)

2.2.2 REMARC

Figure 2.7 shows a block diagram of a microprocessor which includes Reconfigurable Multimedia Array Coprocessor (REMARC) [61]. The RE-MARC consists of a global control unit, coprocessor data registers, and a reconfigurable logic array which includes an 8x8 16-bit processor (nano processor) array. The global control unit controls the execution of the reconfigurable logic array and the transfer of data between the main processor and the reconfigurable logic array through the coprocessor data registers. The MIPS-II ISA is used as the base architecture of the main processor. The MIPS ISA is extended for the REMARC using special instructions. The main processor issues these instructions to the REMARC which executes them in a manner similar to a floating point coprocessor. Unlike a floating point coprocessor,

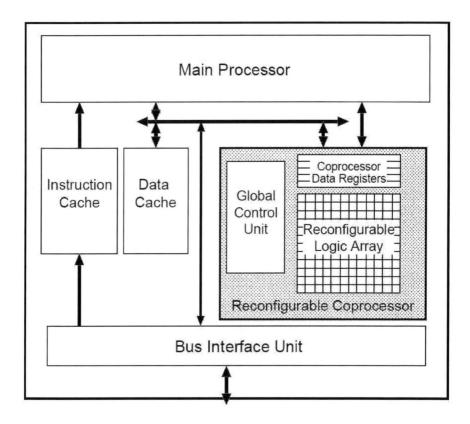

FIGURE 2.7: Block diagram of a microprocessor with REMARC. (From T. Miyamori and K. Olukotun, "A quantitative analysis of reconfigurable coprocessors for multimedia applications." In *Proceedings of IEEE Symposium on FPGAs for Custom Computing Machines*, ⓒ 1998 IEEE. With permission.)

the functions of reconfigurable coprocessor instructions are configurable (or programmable) so that they can be specialized for specific applications.

2.2.2.1 Nano Processor Array Structure

Figure 2.8 shows a block diagram of REMARC. REMARC's reconfigurable logic is composed of an 8x8 array of the 16-bit processors, called nano processors. The execution of each nano processor is controlled by the instructions stored in the local instruction RAM (nano instruction RAM). However, each nano processor does not directly control the instructions it executes. Every cycle the nano processor receives a PC value, "nano PC", from the global control unit. All nano processors use the same nano PC and execute the instructions indexed by the nano PC in their nano instruction RAM. Figure 2.9 shows the architecture of the nano processor. The nano processor contains a 32-entry nano instruction RAM, a 16-bit ALU, a 16-entry data RAM, an

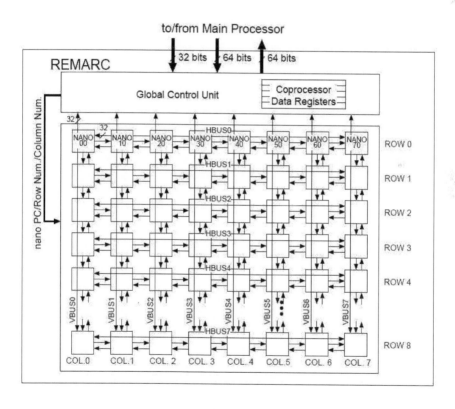

FIGURE 2.8: Block diagram of REMARC. (From T. Miyamori and K. Olukotun, "A quantitative analysis of reconfigurable coprocessors for multimedia applications." In *Proceedings of IEEE Symposium on FPGAs for Custom Computing Machines*, © 1998 IEEE. With permission.)

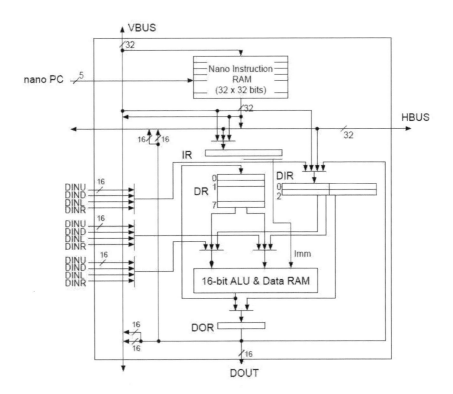

FIGURE 2.9: Block diagram of REMARC. (From T. Miyamori and K. Olukotun, "A quantitative analysis of reconfigurable coprocessors for multimedia applications." In *Proceedings of IEEE Symposium on FPGAs for Custom Computing Machines*, © 1998 IEEE. With permission.)

instruction register (IR), eight 16-bit data registers (DR), four 16-bit data input registers (DIR), and a 16-bit data output register (DOR). Each nano processor can use the DR registers, the DIR registers, and immediate data as the source data of ALU operations. Moreover, it can directly use the DOR registers of the four adjacent nano processors (DINU, DIND, DINL, and DINR) as the source. The nano processors are also connected by the 32-bit Horizontal Buses (HBUSs) and the 32-bit Vertical Buses (VBUSs). Each bus operates as two 16-bit data buses. The 16-bit data in the DOR register can be sent to the upper or lower 16 bits of the VBUS or the HBUS. The HBUSs and the VBUSs allow data to be broadcast to the other nano processors in the same row or column. These buses can reduce the communication overhead between processors separated by long distances. The DIR registers accept inputs from the HBUS, the VBUS, the DOR, or the four adjacent nano processors. Because the width of the HBUS and the VBUS is 32 bits, data on the HBUS or the VBUS are stored into a DIR register pair, DIR0 and DIR1, or DIR2 and DIR3. Using the DIR registers, data can be transferred between nano processors during ALU operations. It takes a half cycle to transfer data using the VBUSs or HBUSs. It should not be a critical path of the design. Other operations, except for data inputs from nearest neighbors, are done within the nano processor. Because the width of a nano processor's datapath is only 16 bits, which is a quarter of those of the general purpose microprocessors, this careful design does not make REMARC a critical path of the chip.

2.2.3 PACT-XPP

Configurable System on Chip architecture [11] consists of a coarse-grained reconfigurable hardware, one LEON micro-controller, and several SRAMtype memory modules as shown in Figure 2.10. The reconfigurable hardware part is the dynamically reconfigurable eXtreme Processing Platform (XPP) from PACT [11]. The XPP architecture realizes a runtime reconfigurable data processing technology that replaces the concept of instruction sequencing by configuration sequencing with high performance application areas envisioned from embedded signal processing to coprocessing in different DSP-like application environments. The main communication bus is chosen to be the AHB from ARM [16]. The size of the XPP architecture will be either 16 ALU-PAEs (4x4-array), or 64 ALU-PAEs (8x8-array), dependent on the application field. To get an efficient coupling of the XPP architecture to AHB, AHB-bridge connects both IO-interfaces on one side of the XPP, input and output interfaces, to the AHB via one module.

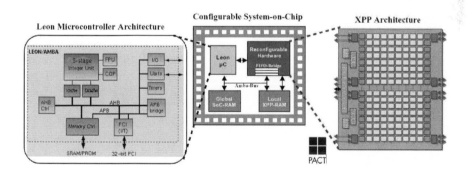

FIGURE 2.10: XPP-based CSoC architecture. (From J. Becker and M. Vorbach, "Architecture, memory and interface technology integration of an industrial/academic configurable system-on-chip (CSoC)." In *Proceedings of IEEE Computer Society Annual Symposium on VLSI*, © 2003 IEEE. With permission.)

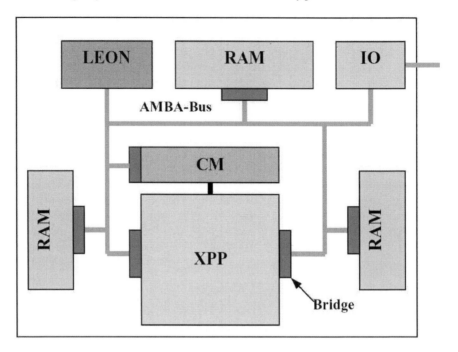

FIGURE 2.11: CSoC RAM topology. (From M. Vorbach, "Reconfigurable processor architectures for mobile phones," In *Proceedings of IPDPS*, © 2003 IEEE. With permission.)

2.2.3.1   Configuration Manager

The Configuration Manager (CM) unit (Figure 2.11) implements a separate memory for faster storing and loading the XPP configurations. If there isn't enough memory space for storing the configurations in local memory, it's possible to use the global CSoC memory to do that. The AHB-bridge for CM will be a single ported SLAVE-AHB-bridge. The transfers of the configurations from global memory to the Configuration Manager will be done by LEON. Therefore the CM have to send a request to LEON and start new configuration transfer.

2.2.3.2   Micro-Controller

The micro-controller on the CSoC is a LEON processor. This processor is a public domain IP core. The LEON VHDL model implements a 32-bit processor conforming to the SPARC V8 architecture. It is designed for embedded applications with the following features on-chip: separate instruction and data caches, hardware multiplier and divider, interrupt controller, two 24-bit timers, two UARTs, power-down function, watchdog, 16-bit I/O port and a flexible memory controller. Additional modules can easily be added using the on-chip AMBA AHB/APB buses. The VHDL model is fully synthesizable with most synthesis tools and can be implemented on both FPGAs and ASICs. The LEON microprocessor acts as a master on our CSoC architecture. The program data for LEON will be transferred via AHB.

2.2.3.3 Processing Array

The XPP architecture is based on a hierarchical array of coarse-grained, adaptive computing elements called Processing Array Elements (PAEs) as shown in Figure 2.12. The strength of the XPP technology originates from the combination of array processing with unique, powerful run-time reconfiguration mechanisms. Since configuration control is distributed over several CMs embedded in the array, PAEs can be configured rapidly in parallel while neighboring PAEs are processing data. Entire applications can be configured and run independently on different parts of the array. Reconfiguration is triggered externally or even by special event signals originating within the array, enabling self-reconfiguring designs. By utilizing protocols implemented in hardware, data and event packets are used to process, generate, decompose and merge streams of data. XPP's main distinguishing features are its automatic packet-handling mechanisms and sophisticated hierarchical configuration protocols. An XPP device contains one or several Processing Array Clusters (PACs), i.e., rectangular blocks of PAEs. Each PAC is attached to a CM responsible for writing configuration data into the configurable objects of the PAC. Multi-PAC devices contain additional CMs for configuration data handling, forming a hierarchical tree of CMs. The root CM is called the su-

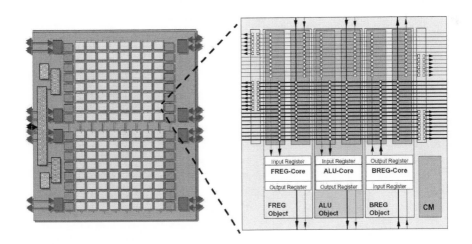

FIGURE 2.12: XPP64 architecture overview and structure of one ALU PAE module. (From M. Vorbach, "Reconfigurable processor architectures for mobile phones," In *Proceedings of IPDPS*, © 2003 IEEE. With permission.)

pervising CM or SCM. The XPP architecture is also designed for cascading multiple devices in a multi-chip. A CM consists of a state machine and internal RAM for configuration caching. The PAC itself contains a configuration bus which connects the CM with PAEs and other configurable objects. Horizontal buses carry data and events. They can be segmented by configurable switch-objects, and connected to PAEs and special I/O objects at the periphery of the device. A PAE is a collection of PAE objects. The typical PAE shown in Figure 2.12 contains a BREG object (back registers)and an FREG object (forward registers) which are used for vertical routing, as well as an ALU object which performs the actual computations. The ALU object's internal structure is shown on the bottom left-hand side of the figure. The ALU implemented performs common fixed-point arithmetical and logical operations as well as several special three-input opcodes like multiply-add, sort, and counters. Events generated by ALU objects depend on ALU results or exceptions, very similar to the state flags of a classical microprocessor. A counter, e.g., generates a special event only after it has terminated. The next section explains how these events are used. Another PAE object implemented in the prototype is a memory object which can be used in FIFO mode or as RAM for lookup tables, intermediate results etc. However, any PAE object functionality can be included in the XPP architecture.

2.2.4   ADRES

Architecture for dynamically reconfigurable embedded systems (ADRES) is a flexible architecture template that includes a tightly coupled very long instruction word (VLIW) processor and a CGRA [60]. Integrating the VLIW processor and the reconfigurable array into a single architecture with two virtual functional views provides advantages over state-of-the-art architectures. A CGRA is intended to efficiently execute only computationally intensive kernels of applications. Therefore, it mostly operates next to a host processor, typically a RISC, that executes the application's remaining parts. Communication between this processor and the configurable accelerator results in programming difficulties and communication overhead that can greatly reduce overall performance. Combining the host processor and the accelerator in one architecture leads to simplified programming and removes the communication bottleneck. Moreover, using a VLIW processor instead of a RISC lets researchers exploit the limited parallelism available in the parts of the code that cannot be mapped to the array, leading to an important overall performance increase. In addition, the CGRA includes components similar to those used in VLIW processors. This creates a resource sharing opportunity that traditional coarse-grained architectures do not extensively exploit. The ADRES architecture template, shown in Figure 2.13, consists of many basic components, including computational, storage, and routing resources. The computational resources are FUs that can execute a set of word-level operations selected by a control signal. Register files (RFs) and memory blocks can

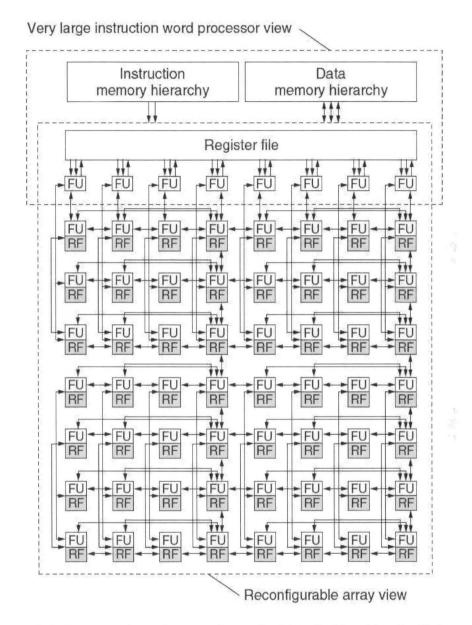

FIGURE 2.13: ADRES core. (From B. Mei, S. Vernalde, D. Verkest, and R. Lauwereins, "Design methodology for a tightly coupled VLIW/reconfigurable matrix architecture: a case study," In *Proceedings of Design Automation and Test in Europe Conference*, © 2004 IEEE. With permission.)

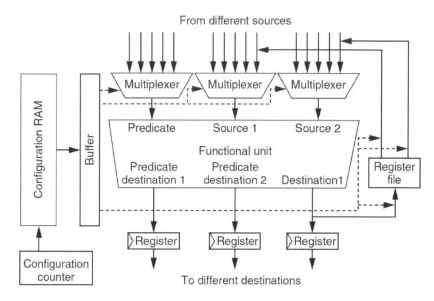

FIGURE 2.14: FU structure. (From B. Mei, S. Vernalde, D. Verkest, and R. Lauwereins, "Design methodology for a tightly coupled VLIW/reconfigurable matrix architecture: a case study," In *Proceedings of Design Automation and Test in Europe Conference*, © 2004 IEEE. With permission.)

store intermediate data. Routing resources include wires, multiplexers, and buses. Basically, the routing resources connect the computational resources and storage resources in a certain topology to form the ADRES array. The whole ADRES array has two functional views: the VLIW processor view and the reconfigurable array view. Both functional views perform tasks exclusively. This makes programming simpler by resource sharing possible, but it doesn't exploit the possible performance gain of parallel processing. The VLIW processor consists of several FUs and a multiport RF, as in typical VLIW architectures, but in this case the VLIW processor also serves as the first row of the reconfigurable array. Some FUs in this first row connect to the memory hierarchy, depending on the number of available ports. Load/store operations available on these FUs facilitate data accesses to the unified architecture's memory. The ADRES array is a flexible template instead of a concrete instance.

Figure 2.14 shows an example of the detailed data path. To remove the control flow inside loops, the FUs support predicated operations. The FUs' results can be written to the distributed RFs, which are small and have fewer ports than the shared RF, or they can be routed to other FUs. To guarantee timing, an output register buffers the FUs' outputs. The multiplexers route data from different sources. The configuration RAM stores a few configurations locally, which can be loaded on a cycle-by-cycle basis. The configurations can also be loaded from the memory hierarchy, at the cost of extra delay if the local configuration RAM isn't big enough. Like instructions in microprocessors, the configurations control the basic components' behavior by selecting operations and controlling multiplexers.

2.2.5   RaPiD

Reconfigurable pipelined datapaths (RaPiDs) are coarse-grained field programmable architectures for constructing deep computational pipelines. As compared to a general purpose processor, a RaPiD can be thought of as a superscalar architecture with hundreds of functional units but with no cache, register file, or crossbar interconnect. Instead of a data cache, data is streamed in directly from external memory or sensors. Instead of an instruction cache, programmed controllers generate a small instruction stream which is decoded as it flows in parallel with the datapath. Instead of a global register file, data and intermediate results are stored locally in registers and small RAMs, close to their destination functional units. Instead of a crossbar, a programmable interconnect is configured to forward data between specific functional units on a per application basis. Removing caches, crossbars, and register files frees up a tremendous amount of area that can be dedicated to compute resources, and reduces the communication delay by shortening wires. Unfortunately, these removals also reduce the types of applications that can be computed on RaPiDs. Highly irregular computations, with complex addressing patterns, little reuse of data, and an absence of fine-grained parallelism will not map well to a RaPiD architecture. However, regular computation-intensive tasks like those found in digital signal processing, scientific computing, graphics, and communications will reap great performance gains on RaPiDs over general purpose processors.

The block diagram in Figure 2.15 breaks down RaPiD into a datapath, a control path, an instruction generator, and a stream manager. The RaPiD datapath is a linear pipeline configured from a linear array of functional units by means of a configurable interconnect. The instruction generator produces a stream which is decoded by the control path. The resulting decoded instructions provide time-varying control for the datapath. The stream manager communicates with external memory (or memory-mapped external sensors) to stream data in and out of the RaPiD datapath.

2.2.5.1   Functional Units

Each functional unit inputs a set of words from the configurable interconnect, performs a computation based on a set of control bits, and outputs results in the form of data words and status bits. The status outputs allow for data-dependent control to be generated. All functional unit outputs pass through a *ConfigDelay* unit which can be configured as 0 to 3 register delays. These optional registers allow for the creation of very deep pipelines. A variety

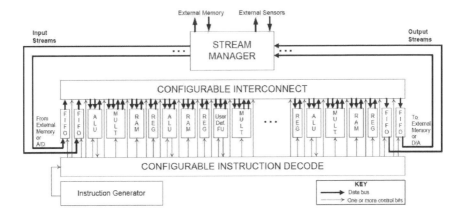

FIGURE 2.15: RaPiD architecture block diagram. (From D. Cronquist, "Architecture design of reconfigurable pipelined datapaths," In *Proceedings of ARVLSI 1999*, © 1999 IEEE. With permission.)

of functional units can be included in a RaPiD architecture. General-purpose functional units like ALUs, multipliers, shifters, and memories are the most common, but for specific domains, a special-purpose functional unit which performs a single function (i.e. has no control inputs) might make the most efficient use of silicon. An example is a Viterbi decoder for communication applications. For other domains, a highly configurable functional unit might be the right choice. For example, a functional unit could be constructed of FPGA-like logic blocks to support a range of bit manipulations like find first one, count ones, and normalize.

2.2.5.2   Configurable Interconnect

The configurable interconnect consists of a set of T segmented tracks that run the entire length of the datapath. Each track contains a set of bus segments, some of which are connected by bus connectors—configurable connections that can be open or up to three register delays. All buses have the same width, which matches the data width operated on by the functional units. Some functional units may require or produce double-width data values, which are communicated via two buses. These values can be treated as two independent single-width values and routed independently, for example, to two different ALUs for double-precision accumulation. An input to a functional unit can be zero (GND) or any one of the T tracks from the interconnect. To accomplish this, each data input is driven by a $(T +1) : 1$ multiplexer, whose $[\lg(T +1)]$ select lines are driven by control signals. The zero input can be used, for example, to clear registers. RaPiD allows each functional unit output to drive an arbitrary number of buses via T tristate drivers which are configured using T control bits. Since each tristate driver is configured independently, an output can fan out to several buses or none at all if the functional unit is unused.

2.2.6 PipeRench

PipeRench [14, 74] is a coarse-grained reconfigurable fabric—an interconnected network of configurable logic and storage elements. By virtualizing the hardware, PipeRench efficiently handles computations. Using a technique called pipeline reconfiguration, PipeRench improves compilation time, reconfiguration time, and forward compatibility. PipeRench's architectural parameters (including logic block granularity) optimize the performance of a suite of kernels, balancing the compiler's needs against the constraints of deep-submicron process technology. PipeRench is particularly suitable for stream-based media applications or any applications that rely on simple, regular computations on large sets of small data elements.

Figure 2.16a is an abstract view of the PipeRench architectural class, and Figure 2.16b is a more detailed view of a processing element (PE). PipeRench contains a set of physical pipeline stages called stripes. Each stripe has an interconnection network and a set of PEs. Each PE contains an arithmetic logic unit and a pass register file. Each ALU contains lookup tables (LUTs)and extra circuitry for carry chains, zero detection, and so on. Designers implement combinational logic using a set of NB-bit-wide ALUs. The ALU operation is static while a particular virtual stripe resides in a physical stripe. Designers can cascade the carry lines of PipeRench's ALUs to construct wider ALUs, and chain PEs together via the interconnection network to build complex combinational functions. Through the interconnection network, PEs can access operands from registered outputs of the previous stripe, as well as registered or unregistered outputs of the other PEs in the same stripe. Because of hardware virtualization constraints, no buses can connect consecutive stripes. However, the PEs access global I/O buses. These buses are necessary because an application's pipeline stages may physically reside in any of the fabric's stripes. Inputs to and outputs from the application must use a global bus to get to their destination. The pass register file provides efficient pipelined interstripe connections. A program can write the ALU's output to any of the P registers in the pass register file. If the ALU does not write to a particular register, that register's value will come from the value in the previous stripe's corresponding pass register. The pass register file provides a pipelined interconnection from a PE in one stripe to the corresponding PE in subsequent stripes. For data values to move laterally within a stripe, they must use the interconnection network (see Figure 2a). In each stripe, the interconnection network accepts inputs from each PE in that stripe, plus one of the register values from each register file in the previous stripe. Moreover, a barrel shifter in each PE shifts its inputs B . 1 bits to the left (see Figure 2.16b). Thus, PipeRench can handle the data alignments necessary for word-based arithmetic.

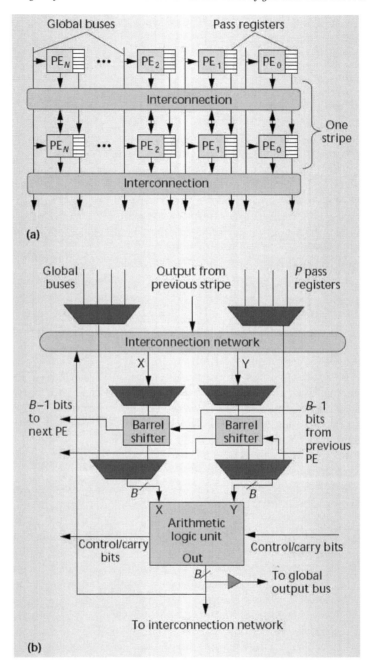

FIGURE 2.16: PipeRench architecture: (a) each stripe contains processing elements (PEs) and an interconnection; (b) a detailed view of a PE and its connections. (From S.C. Goldstein, H. Schmit, M. Budiu, S. Cadambi, M. Moe, and R.R. Taylor, "PipeRench: a reconfigurable architecture and compiler," *IEEE Computer*, © 2000 IEEE. With permission.)

2.3 Design Space Exploration

Coarse-grained architectures can be tailored and optimized for different application domains. The goal is to design a domain-specific processor that provides just enough flexibility for that domain while minimizing the energy consumption for a given level of performance. Achieving this goal requires numerous architectural choices and finding the optimal architecture involves many trade-offs in choosing values for each parameter. To support the finding of a suitable CGRA for a given application domain, the architecture exploration flows have been suggested in [7, 8, 21, 31, 38, 54–56, 60, 63]. This section presents such environments and architecture explorations of some representative CGRA examples.

2.3.1 PACT-XPP - PARO

Design space exploration of PACT-XPP is achieved by using existing mapping methodology PARO [28] when generating synthesizable descriptions of massively parallel processor arrays from regular algorithms. The design flow of PARO is depicted in Figure 2.17. Starting from a given nested loop program in a sequential high-level language (subset of C) the program is parallelized by data dependence analysis into single assignment code (SAC). This algorithm class can be written as a class of recurrence equations. With this representation of equations and index spaces several combinations of parallelizing transformations in the polytope model can be applied.

- *Affine Transformations*, like skewing of data spaces or the embedding of variables into a common index space.

- *Localization* of affine data dependencies to uniform data dependencies by propagation of variables from one index point to a neighbor index point.

- *Operator Splitting*, equations with more than one operation can be split into equations with only two operands.

- *Exploration of Space-Time Mappings*. Linear transformations are used as space-time mappings in order to assign a processor p(space) and sequencing index t(time) to index vectors. The main reasons for using linear allocation and scheduling functions is that the data flow between PEs is local and regular which is essential for low-power VLSI implementations.

- *Partitioning*. In order to match resource constraints such as limited number of processing elements, partitioning techniques have to be applied.

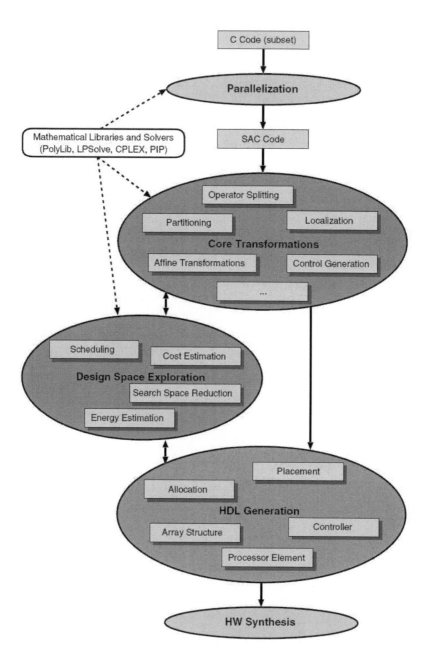

FIGURE 2.17: PARO design flow. (From F. Hannig, H. Dutta, and J. Teich, "Mapping of regular nested loop programs to coarse-grained reconfigurable arrays—constraints and methodology," In *Proceedings of IEEE International Parallel and Distributed Processing Symp*, © 2004 IEEE. With permission.)

- *Control Generation.* If the functionality of one processing element can change over time, control mechanisms are necessary. Further control structures are necessary to control the internal schedule of a PE.

- *HDL Generation & Synthesis.* Finally after all the refining transformations a synthesizable description in a hardware description language like VHDL may be generated. This is done by generating one PE and the repetitive generation of the entire array.

2.3.2   KressArray Xplorer

An architecture of the KressArray family is a regular array of coarse-grain reconfigurable Data Path Units (rDPUs) [30] of multiple-bit data path width. To optimize a KressArray architecture for a particular application area, the functionality of the rDPUs can be tailored to the demands of this application domain. Often, a tradeoff has to be found between the silicon area and the performance of the operators, resulting in different implementations of a certain operator. To assist the user in finding the best suitable architecture for a given application domain, an interactive design space exploration environment called KressArray Xplorer [31] has been implemented at Kaiserslautern University. The Xplorer is implemented in C/C++ and is running on UNIX and LINUX. The KressArray Xplorer is based on the MA-DPSS (Multi-Architecture Datapath Synthesis System) design framework for KressArrays, which is used to map datapaths described in the high-level ALE-X language onto a KressArray. ALE-X statements may contain arithmetic and logic expressions, conditions, and loops, that evaluate iterative computations on a small number of input data.

An overview of the KressArray Xplorer is given in Figure 2.18. The user provides a description of the application using the high-level ALE-X language. The ALE-X compiler creates an intermediate format, which is used by all other tools. At the beginning of the exploration process, the minimal requirements for the architecture are estimated and added to the intermediate format. Then, the user can perform consecutive design cycles, using the mapper and the data scheduler. Both tools generate statistical data, which is evaluated by an analyzer to make suggestions for possible architecture enhancements, which are presented to the user by an interactive editor. This editor is also used to control the design process itself. When a suitable architecture has been found, a HDL description (currently Verilog) can be generated from the mapping for simulation.

2.3.3   ADRES - DRESC

For ADRES architecture exploration, its own compiler and architecture description framework are used [60]—it is called the dynamically reconfigurable embedded system compiler (DRESC). Figure 2.19 shows the compiler

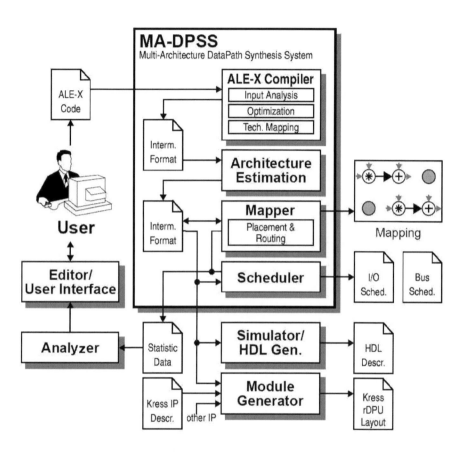

FIGURE 2.18: KressArray Xplorer overview. (From R. Hartenstein, M. Herz, T. Hoffmann, and U. Nageldinger, "KressArray Xplorer: a new CAD environment to optimize reconfigurable datapath array architectures," In *Proceedings of Asia and South Pacific Design Automation Conference*, © 2000 IEEE. With permission.)

framework. A design starts from a C-language description of the application. On the basis of execution time and possible speedup, the profiling and partitioning step identifies the candidate computation-intensive loops (kernels) for mapping onto the reconfigurable array. Source-level transformations let us pipeline the kernel software and maximize performance. In the next step, they use Impact, a VLIW compiler framework, to parse the C code and perform analysis and optimization. Impact emits an intermediate representation called Lcode, which serves as input for scheduling. Applying a novel modulo scheduling algorithm that takes program and architecture representation as input achieves high parallelism for the kernels, whereas applying traditional instruction-level parallelism (ILP) scheduling techniques reveals the available moderate parallelism for the nonkernel code. Our tools automatically identify and handle communication between these two parts and generate scheduled code for both the reconfigurable array and the VLIW processor. The tools also generate a cosimulator, using the architecture description and the scheduled code as inputs, and designers can use this simulator to obtain quality metrics for the architecture instance under test.

As the right-hand side of Figure 2.19 shows, an XML-based language describes the target architecture. This architecture's high-level parameterized description lets a designer quickly specify different architecture variations. The parser and abstraction steps transform the architecture into an internal, more detailed, graph representation.

2.4   Code Compilation and Mapping

Coarse-grained reconfigurable architectures can enhance the performance of critical loops and computation-intensive functions. Such architectures need efficient compilation techniques to map algorithms onto customized architectural configurations and some research works have been presented which deal with the compilation to coarse-grained reconfigurable architectures [1, 28, 29, 39, 50–53, 60, 66, 69, 77, 78, 80, 81]. The objective of such works is to analyze each loop, perform optimizations, and generate a complete, efficient execution schedule that specifies the temporal ordering of each operation, where on the PE Array each of the operations will execute. In this section, we introduce two representative compilation approaches for CGRAs—MorphoSys and ADRES.

2.4.1   MorphoSys - Hierarchical Loop Synthesis

In [77], authors describe a compiler framework to analyze SA-C programs, perform optimizations, and automatically map the application onto the MorphoSys [75]. The mapping process is static and it involves operation schedul-

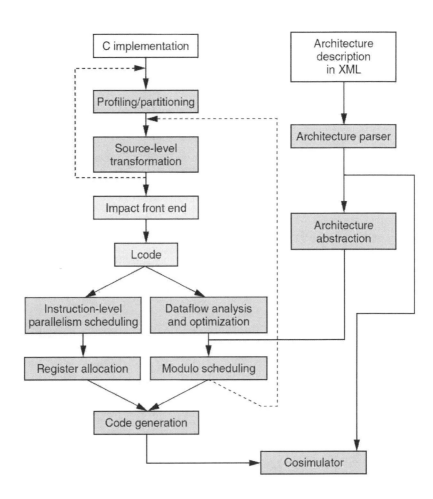

FIGURE 2.19: Dynamically reconfigurable embedded system compiler (DRESC) framework. (From B. Mei, S. Vernalde, D. Verkest, and R. Lauwereins, "Design methodology for a tightly coupled VLIW/reconfigurable matrix architecture: a case study," In *Proceedings of Design Automation and Test in Europe Conference*, © 2004 IEEE. With permission.)

ing, processor allocation and binding, and register allocation in the context of the MorphoSys architecture. The compiler also handles issues concerning data streaming and caching in order to minimize data transfer overhead. Figure 2.20 shows the flow of compilation. The right-side branch of compilation after code partitioning represents the compilation of code that is not within loops. This phase of code generation is, essentially, similar to that of traditional compilers. The left-hand branch represents the heart of this compiler. The process of generating the detailed execution schedule is referred to as "loop synthesis."

2.4.1.1 Function Inlining

Since SA-C does not support pointers or recursion, every function in SA-C can be inlined. Inlining a function ensures that the context within which a function is called is exposed. This makes a difference during code partitioning, as a particular function can be called either from within a loop or from outside a loop, and this will determine whether the particular function will be mapped to the RC Array or to the Tiny RISC processor.

2.4.1.2 Transformation to Context Codes

The loops in the SA-C program are mapped onto the RC Array for execution. Hence, as a requirement, every simple node within a loop must have a one-to-one correspondence to an RC Array Context code. Most of the operations within a loop will usually correspond directly to an RC Array context code. However, at times, the operation is implicit and may be associated with a group of graph nodes. During the optimization phase, the compiler essentially performs a pattern-matching pass to find such candidate groups of nodes that can represent a single RC Array context code, and transforms such nodes. On the other hand, there may be certain nodes that do not directly correspond to any of the RC Array context codes. These operations can, however, be represented as a sequence of context codes that have the same effect. In such cases, the operation execution latencies of these nodes are updated to reflect the time required to execute this sequence of contexts. Ordinarily, for all other operations that directly correspond to an RC Array context code, the execution latency of the operation is 1 clock cycle.

2.4.1.3 Hierarchical Loop Synthesis

All code in the SA-C program is statically scheduled for execution by the compiler. The compiler adopts a hierarchical approach to solve the problem of mapping SA-C loops. Loops are synthesized based on their relative position in the loop hierarchy, with the innermost loop defined to be at the bottom of the loop hierarchy. The compiler's approach is to synthesize the innermost loop, and then progressively move up the loop hierarchy until the outermost loop is

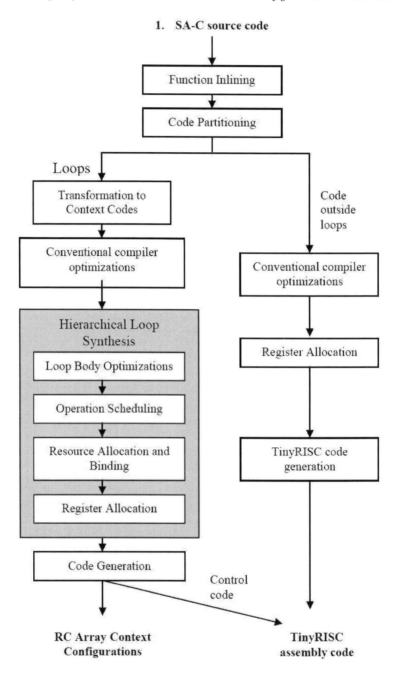

FIGURE 2.20: Flow of compilation. (From G. Venkataramani, W. Najjar, F. Kurdahi, N. Bagherzadeh, and W. Bohm, "A compiler framework for mapping applications to a coarse-grained reconfigurable computer architecture," In *Proceedings of International Conference on Compilers, Architecture, and Synthesis for Embedded Systems*, © 2001 IEEE. With permission.)

synthesized. The compiler framework defines different execution models based on the loop's generator.

2.4.2 ADRES - Modulo Scheduling

As mentioned in 2.3.3, DRESC is the compiler framework of ADRES and the compiler's core technology is the modulo scheduling algorithm, which can map loops onto the ADRES architecture in a highly parallel way [60]. The objective of modulo scheduling, a widely used software pipelining technique, is to execute multiple iterations of the same loop in parallel. To achieve this, the compiler constructs a schedule for one loop iteration such that this same schedule repeats at regular intervals with respect to intra and inter-iteration dependences and resource constraints. This initiation interval shows how many cycles elapse before the next iteration of the loop starts executing. Because the initiation interval is inversely proportional to performance, it reveals the scheduled loop's performance. Applied to coarse-grained architectures, modulo scheduling complexity increases drastically because it must combine three subproblems—placement, routing, and scheduling—in a modulo-constrained space.

In Figure 2.21a, a simple data dependence graph, representing a loop body and a 2x2 array, illustrates the problem. Figure 2.21b shows the scheduled loop, which is a space-time representation of the scheduling space. (The 2x2 array is flattened to 1x4 for convenience of drawing.) The dashed lines represent routing possibilities between the FUs. Placement determines which of a 2D array's FUs to place an operation on, and scheduling determines the cycle in which to execute that operation. Routing connects the placed and scheduled operations according to their data dependences. The schedule on the 2x2 array appears in Figure 2.21c, where the initiation interval equals 1. FUs 1 through 4 are configured to execute operations n2, n4, n1, and n3, respectively. Overlapping a loop's different iterations achieves four instructions per cycle (IPC) in this simple example.

Applying a novel modulo scheduling algorithm that takes program and architecture representation as input achieves high parallelism for the kernels, whereas applying traditional instruction-level parallelism (ILP) scheduling techniques reveals the available moderate parallelism for the nonkernel code. DRESC automatically identify and handle communication between these two parts and generate scheduled code for both the reconfigurable array and the VLIW processor. The tools also generate a cosimulator, using the architecture description and the scheduled code as inputs, and designers can use this simulator to obtain quality metrics for the architecture instance under test.

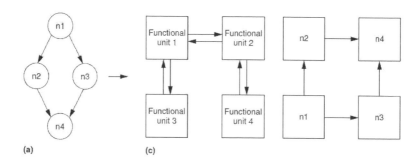

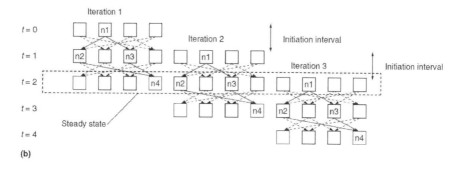

FIGURE 2.21: Mapping a kernel onto a 2x2 reconfigurable array: (a) data dependence graph of the kernel; (b) schedule with the initiation interval = 1; (c) resulting configuration. (From B. Mei, "Architecture exploration for a reconfigurable architecture template," *IEEE Design & Test of Computers,* © 2005 IEEE. With permission.)

2.5 Physical Implementation

Many reconfigurable architectures have been implemented with various technologies [9, 11, 14, 37, 39, 46, 49, 65, 74, 78]. Most of these researches aim to analyze hardware cost of CGRA such as area, delay or power at gate-level. In this section, we describe physical implementation cases of some CGRAs and analysis about hardware cost of the CGRAs.

2.5.1 MorphoSys

The first implementation of MorphoSys is called the M1 chip [58]. M1 has been designed for operation at 100 MHz clock frequency, using a 0.35 μm 4 layers of metal CMOS technology. The complete design methodology includes two fundamental procedures: standard cell approach and custom design approach. At first, the complete MorphoSys reconfigurable system was modeled using behavioral VHDL. Then, the TinyRISC and DMA controller were re-modeled to structural VHDL, and the EDIF net-list, together with the constraints file, is generated from VHDL using Synopsys Design Compiler. The original EDIF netlist is then converted to UPR file using EDIF2UPR procedure. Then Mentor Graphics AutoCells in GDT package was employed for layout synthesis using UPR file and constraints file. The SDF (Standard Delay File) from AutoCells is sent back to Design Compiler to reoptimize the design. The other components were completely custom-designed (including TinyRISC's data register file and data cache core) using Magic 6.5. Routing of symmetrical RC Array is done manually using L language. By taking both layouts from AutoCells and Magic, Mentor Graphics MicroPlan and Micro-Route were employed to handle the integration of the individual components. The final layout is illustrated in Figure 2.22. Its dimensions are 14.5mm by 12.5mm.

2.5.2 PACT-XPP

Figure 2.23 shows synthesis layout of XPP/LEON-based CSoC architecture [11]. The LEON processor architecture, illustrated in section 2.2.5.2, has been first synthesized onto UMC 0.18 μm technology. The LEON processor core needs 1.8 mm^2 and can be clocked up to 200 Mhz. The synthesis was done hierarchically (synthesized netlists and place&route performed separately for different LEON modules) and completely flattened. The major advantage in using flattened synthesis was the better performance for the critical path of LEON in its Integer Unit. XPP ALU-PAEs have been synthesized in 0.13 μm CMOS technology in different synthesis strategies: hierarchical, semi-hierarchical, and flattened. The semi-hierarchical strategy gives the best

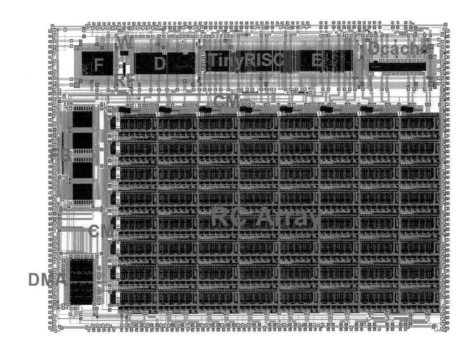

FIGURE 2.22: Layout of MorphoSys M1 chip.

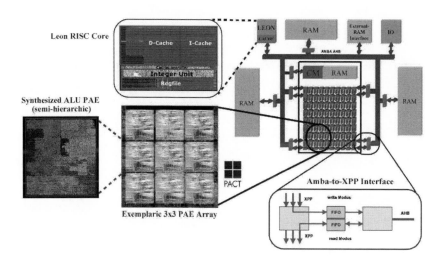

FIGURE 2.23: XPP/LEON-based CSoC architecture with multi-layer AMBA interface. (From J. Becker, "An industrial/academic Configurable System-on-Chip project (CSoC): coarse-grain XPP-/Leon-based architecture integration," In *Proceedings of IEEE*, © 2003 IEEE. With permission.)

performance/area trade-offs, still allowing a parametrizable modular design flow (see Figure 2.23).

2.5.3 ADRES

The Reconfigurable Cell(RC) described in the Section 2.2.4 was custom implemented with the Infineon Technologies CMOS 130nm 6 metals process technology [78]. The methodology was to start with standard cells. Placement was done manually and routing was partially automatic, partially manual. The layout implementation is based on standard cells. Medium drive strength is used inside a subblock and strong buffering for the lines, which broadcast inside the RC. The cells mostly are hand-placed, because many parts of the microarchitecture are regular. Routing inside a subblock is done automatically. The routing between the subblocks is done manually. The total area of the layout is 0.196 mm^2. The contributions of the subblocks is as follows: configuration block 50%, external interfaces for input and output 6%, registerfile 9% and ALU 19%. About 15% of the area of an RC are consumed by the interconnect between the subblocks. The routing of the gates inside the subblocks consumes metal 2 and 3. The routing between the subblocks can use metal 1 up to metal 4. The metal 4 layer was required for traversing the three functional units inside the ALU. Figure 2.24 shows floorplanning of a RC.

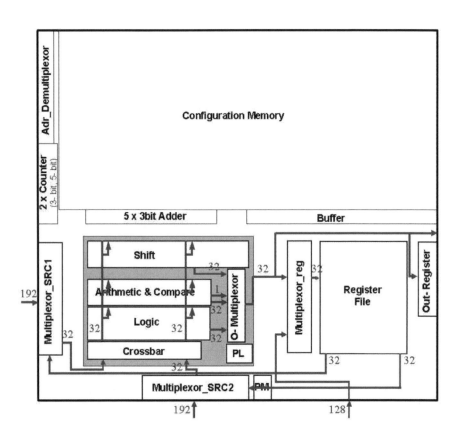

FIGURE 2.24: Floorplan of the reconfigurable cell in ADRES. (From F. Vererdas, M. Scheppler, W. Moffat, and B. Mei, "Custom implementation of the coarse-grained reconfigurable ADRES architecture for multimedia purposes," In *Proceedings of International Conference on Field Programmable Logic and Applications*, © 2005 IEEE. With permission.)

2.5.4 PipeRench

[74] describes the implementation of PipeRench in a six-metal layer 0.18 micron CMOS process. The total die area is 7.3x7.6 mm². Transistor count is 3.65 million. The chip has 132 pins, which includes a 72-pin data interface, 5-bit test interface and 53 pins for power and ground. There are 3.3V and 1.8V supplies for the I/Os and core, respectively. The core area is divided into two areas: (1) the fabric, and (2) the virtualization and interface logic. The fabric consists of sixteen stripes.

The virtualization and interface logic is implemented using standard cells. The configuration data is stored in 22 SRAMs, each with 256 32-bit words. Four 256 word by 32-bit dual-port SRAMs are used for storage of state information in the fabric. Sixteen dual-port SRAMs, (32 x 16 bits), are used to queue data between the interface and fabric. The virtualization storage and logic consumes less area than the 16 stripe fabric and stores 256 virtual stripes. This implementation can virtualize a hardware design that is sixteen times its own size. The chip has two clock inputs: one clock controls the operation of the fabric and virtualization; the second clock controls the off-chip interface. These clocks are fully decoupled; all data transactions across the clock domains go through the memory queues between the interface and fabric and all control signals pass through a synchronizer. The fabric clock is designed to operate at 120 MHz under worst-case voltage and temperature conditions. The interface clock is designed to operate at 60 MHz.

The layout of a PE is shown in Figure 2.25 with the top four layers of metal transparent and with some basic components labeled. The dimensions of this cell are 325 mm by 225 mm. Sixteen of these PEs compose a stripe, and the fabric in Figure 2.26 contains 256 of these PEs. The PE area is dominated by interconnect resources such as multiplexors and bus drivers. The transistor density of this layout is not dense. The dimensions of the PE layout are dictated by the interconnect to other PEs in the stripe, and by the global buses, which run vertically over the PE cell.

2.6 Summary

This chapter presents the CGRA research trends in recent years. Some representative CGRAs are introduced in four aspects: architecture, design space exploration (DSE), code compilation & mapping and physical implementation. First of all, we describe entire structure, PE array fabric and memory organization of the CGRA examples in the architecture aspect. In the design space exploration, we introduce architecture optimization issues and existing DSEs to generate efficient CGRA structures. Code compilation & mapping aspect describes noted automatic compilation flows and mapping algorithms

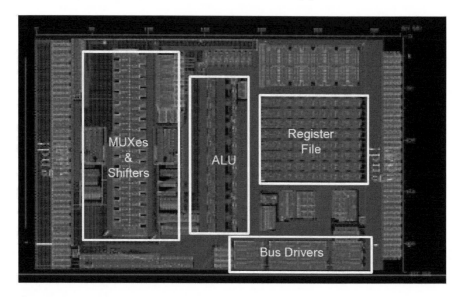

FIGURE 2.25: PipeRench PE floorplan. (From H. Schmit, D. Whelihan, M. Moe, A. Tsai, B. Levine, and R. Taylor, "PipeRench: A Virtualized programmable datapath in 0.18 micron technology," In *Proceedings of IEEE Custom Integrated Circuits Conference*, © 2002 IEEE. With permission.)

of existing CGRAs. Finally, we illustrate physical implementation examples of CGRAs to show analysis of their hardware costs.

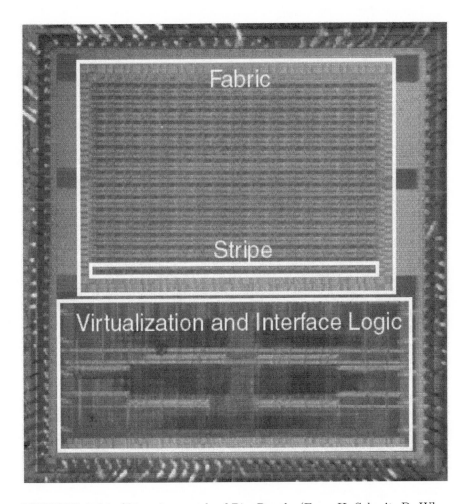

FIGURE 2.26: Chip micrograph of PipeRench. (From H. Schmit, D. Whe-lihan, M. Moe, A. Tsai, B. Levine, and R. Taylor, "PipeRench: A virtualized programmable datapath in 0.18 micron technology," In *Proceedings of IEEE Custom Integrated Circuits Conference*, © 2002 IEEE. With permission.)

Chapter 3

CGRA for High Performance and Flexibility

3.1 Performance versus Flexibility for Embedded Systems

Figure 3.1 shows the general relationship between performance and flexibility for different kinds of IP-types for embedded systems. The more flexible a IP is, the lower performance it has. Therefore, a traditional hardware-software codesign flow maps applications into application-specific integrated circuit (ASIC) or general purpose processor (GPP). Performance-critical parts in an application software are mapped to ASIC and the rest of the application is mapped to GPP. Such a codesign flow enables the embedded systems to be optimized for the specific applications. However, the embedded applications rapidly change by consumers' need and it means the embedded systems should be re-fabricated according to the changed applications in order to meet contraints on performance and power. Therefore, such a traditional codesign method exposes its limitation in the aspect of flexibility.

To overcome such a limitation, coarse-grained reconfigurable architecture (CGRA) has emerged as a suitable solution for the embedded systems. The incompatible demands (high performance and flexibility) can be compromised by CGRA because CGRAs are implemented by domain-specific design method. The domain-specific design method means that a system is designed and optimized for supporting application-groups composed of several applications. Common examples for such applications running on the embedded systems are digital signal processing (DSP) applications like audio signal processing, audio compression, digital image processing, video compression, speech processing, speech recognition, and digital communications. Such applications have many subtasks such as trigonometric functions, filters and matrix/vector operations and the subtasks show computation-intensive and data-parallel characteristics with high regularity. Therefore, in order to support these characteristics of the applications, CGRA is composed of reconfigurable ALU-like datapath blocks processing word data and supports dynamic reconfiguration of interconnections among the processing blocks. Therefore, CGRA not only boosts the performance by efficiently extracting high computational density

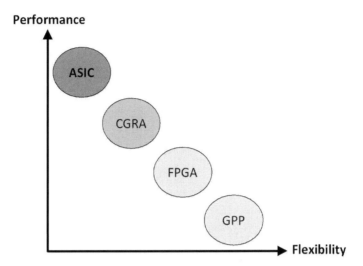

FIGURE 3.1: Performance and flexibility tradeoffs for different kinds of IP-types.

on coarse-grained datapath array but it can also be reconfigured to adapt different characteristics of the applications. FPGA can be also considered as a solution because of its flexibility but it shows very lower performance level and huge hardware cost compared with ASIC—conventional FPGA has logic blocks handling bit data and requires long reconfiguration time with huge routing area overhead.

Therefore, CGRA enables design space tradeoffs between performance and flexibility as depicted in Figure 3.1. Since CGRA provides wider applicability than an ASIC, with performance better than that of general purpose processors, it presents the system designer with the possibility of making tradeoffs between performance and range of applications desired. These tradeoffs present economic benefit to the designer since the number of ASICs (or the ASIC die area) required can be reduced through the use of CGRAs without sacrificing performance levels.

3.2   Performance Evaluation Examples

Many kinds of coarse-grained reconfigurable architecture have been proposed and designed with the increasing interests in reconfigurable computing in recent years [30]. In this section, we introduce performance evaluation examples of some CGRAs with experimental results showing speeds up to several times faster than fixed processors for specific application domains.

3.2.1 MorphoSys

The MorphoSys [75] is a representative example of a coarse-grained reconfigurable architecture. It is an integrated system-on-chip targeted at data-parallel applications with high throughput requirements. The reconfigurable element of MorphoSys is an 8x8 array of processing elements that support a SIMD computational model. The MorphoSys has been evaluated as a platform for many applications such as multimedia, wireless communication, signal processing, and computer graphics. Among them, we describe four cases of performance comparison among the MophoSys and other conventional IPs—two major functions (motion estimation and DCT) required of the MPEG video encoder, important target recognition application (Automatic Target Recognition), and data encryption algorithms. Motion estimation and DCT function have a high degree of data-parallelism and tight real-time constraints. Automatic Target Recognition (ATR) is one of the most computation-intensive applications. The International Data Encryption Algorithm (IDEA) for data encryption is typical of data-intensive applications.

3.2.1.1 Motion Estimation for MPEG

Motion estimation is widely adopted in video compression to identify redundancy between frames. The most popular technique for motion estimation is the block matching algorithm because of its simple hardware implementation. Some standards also recommend this algorithm. Among the different block-matching methods, Full Search Block Matching (FSBM) gives optimal solution with low control overhead and the maximum computation. Figure 3.2 shows performance comparison for matching one 8x8 reference block against its search area of eight pixels displacement. The ASIC architectures (ASIC1 [33], ASIC2 [64] and ASIC3 [79]) employ customized hardware units such as parallel adders to enhance performance. A high performance DSP processor, TMS320C64X [5] requires 2,100 cycles for performing the same computation. The number of processing cycles for MorphoSys is even less than the number of cycles required by the ASIC designs. Since MorphoSys is not an ASIC, its performance with regard to these ASICs is significant. Using the same parameters above, Pentium MMX [5]takes almost 29,000 cycles for the same task. When scaled for clock speed and same technology (fastest Pentium MMX fabricated with 0.35 micron technology operates at 233 MHz, therefore the cycles are divided by 2.33 as MorphoSys operates at 100 MHz), this amounts to more than 20X difference in performance.

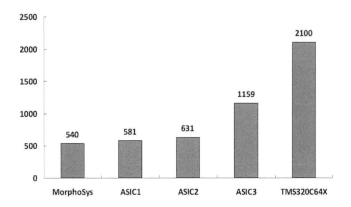

FIGURE 3.2: Performance comparison (in cycles) for motion estimation. (From H. Singh, "MorphoSys: An integrated reconfigurable system for data-parallel and computation-intensive applications. *IEEE Transactions on Computers*, © 2000 IEEE. With permission.)

3.2.1.2   Discrete Cosine Transform (DCT) for MPEG

The forward and inverse DCT are used in MPEG encoders and decoders. The DCT and IDCT are applied on a 2D image pixel block. It is possible to implement the 2D DCT (and IDCT) using 1D DCT on the rows and then the columns (or vice versa) of the image block, using the separability property. The relative performance of MorphoSys and other processors is illustrated in Figure 3.3. MorphoSys requires 37 cycles to complete 2D DCT (or IDCT) on a 8x8 block of pixel data. The reconfigurable coprocessor, REMARC [61] takes 54 cycles. A high performance DSP video processor, TMS320C64X [5] needs 92 cycles, while a dedicated superscalar multimedia processor, the V830R/AV [3], requires 201 cycles. However, Pentium II [22] uses 240 cycles to compute the 2D DCT/IDCT with manually optimized code using MMX instructions. If performance of Pentium II is scaled for same fabrication technology (0.35 micron) as MorphoSys, the former still needs 103 cycles for this computation.

3.2.1.3   Automatic Target Recognition

Automatic Target Recognition (ATR) is the machine function of automatically detecting, classifying, recognizing, and identifying an object. The computation levels reach the hundreds-of-teraops range when targets are partially obscured. Even though there are many algorithmic choices available to implement an ATR system, the system parameters implemented in [13] are chosen for performance comparison. The Automatic Target Recognition (ATR) systems from [13] and [2] are used for comparison. Two Xilinx XC4013 FPGAs are used in the Mojave system [13]. The Splash system consisting of 16 Xilinx XC4010 FPGAs, is used in [2]. The image size of the chips is 128 x 128 pixels

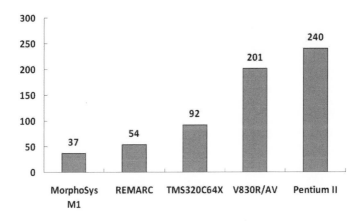

FIGURE 3.3: DCT/IDCT performance comparison (cycles). (From H. Singh, "MorphoSys: An integrated reconfigurable system for data-parallel and computation-intensive applications. *IEEE Transactions on Computers*, © 2000 IEEE. With permission.)

and the template size is 8x8 bits. The complete Second Level of Detection (SLD) processing time for one pair of templates is 6.0 ms for MorphoSys. About 24 ms are required for the Mojave system and each processing element (eight FPGAs) of the Splash 2 system [2] perform the computation in 12ms. These results are shown in Figure 3.4. The operating frequency of the FPGA-based Mojave and Splash 2 systems is about 20MHz, but this is relatively fast for the FPGA domain. The frequency difference is also offset by the fact that ATR operations are inherently fine-grain and involve bit-level computations, which map very well on FPGAs. Even though MorphoSys is a coarse-grained system, it achieves better performance than the above FPGA-based systems for the bit-level ATR operations.

3.2.1.4 International Data Encryption Algorithm

The International Data Encryption Algorithm (IDEA) is a typical example of computation-intensive and data-parallel application domain. IDEA involves processing of plaintext data (data to be encrypted) in 64-bit blocks with a 128-bit encryption/decryption key. Figure 3.5 shows the scaled relative performance of MorphoSys and other IP-cores. A software implementation of the IDEA on a Pentium II processor requires 357 clock cycles to generate on ciphertext block (performance profiled using Intel Vtune). An ASIC, HiPCrypto [73], that provides a dedicated hardware implementation of IDEA produces seven ciphertext data every 56 cycles. The performance of these two systems is scaled to the 100 MHz operating frequency of MorphoSys, resulting in 153 effective cycles for Pentium II (normally operating at 233 MHz for 0.35

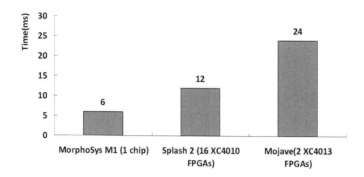

FIGURE 3.4: Performance comparison of MorphoSys for SLD (ATR). (From H. Singh, "MorphoSys: An integrated reconfigurable system for data-parallel and computation-intensive applications. *IEEE Transactions on Computers*, © 2000 IEEE. With permission.)

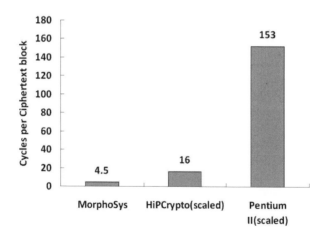

FIGURE 3.5: Performance comparison for IDEA mapping on MorphoSys. (From H. Singh, "MorphoSys: An integrated reconfigurable system for data-parallel and computation-intensive applications. *IEEE Transactions on Computers*, © 2000 IEEE. With permission.)

micron fabrication technology) and 16 effective cycles for HiPCrypto (which operates at 53 MHz).

3.2.2   ADRES

The ADRES [60] is a flexible coarse-grained reconfigurable architecture template that tightly couples a VLIW processor and a coarse-grained reconfigurable array. The reconfigurable array is used to accelerate the dataflow-like kernels in a highly parallel way, whereas the VLIW executes the non-kernel code by exploiting instruction-level parallelism (ILP). In this section, we introduce two cases of application-mapping on the ADRES—H.264/AVC decoder and image processing algorithm.

3.2.2.1   H.264/AVC Decoder

[59] describes the process and results of mapping H.264/AVC decoder onto the ADRES architecture. The authors instantiate an instance from the ADRES architecture to conduct the experiments. It is an 8x8 array that consists of 64 FUs, which include 16 multipliers, 40 ALUs and 8 load/store units. Each FU is not only connected to 4 nearest neighbor FUs, but also 4 FUs with one hop away along the vertical and horizontal directions. There are also 56 distributed RFs and a VLIWRF. A distributed RF is connected to the FU next to it and the FUs in diagonal directions. The scheduling results for kernels are listed in Figure 3.6. The second column of Figure 3.6 is the number of operations of the loop body. Initiation interval (II) means a new iteration can start at every II cycle. It is also equivalent to the number of configuration contexts needed for this kernel. Instruction per-cycle (IPC) reflects the parallelism. Stages refer to total pipeline stages which have an impact on prologue/epilogue overhead. Scheduling time is the CPU time to compute the schedule on a Pentium M 1.4GHz/Linux PC. To verify the complete application, they use a co-simulator to simulate the compiled application using several bitstreams. All these bitstreams are CIF format (352x288). The simulation results are listed in Figure 3.7. *Foreman* is a simple bitstream with less motion, whereas *mobile* is much more complex and has more motion. *bf* and *nobf* indicate whether there are B-frames or not. The last part of the bitstream names indicates the bit-rate at 25 frames/sec. The second and the third columns are the results of performing H.264/AVC on the VLIW and the ADRES architecture respectively; the last column shows the results of speed-up (kernel/overall) over the VLIW, respectively. The results show for they can achieve 4.2 times of kernel speed-up. The overall speed-up is around 1.9 times for low bit-rate bitstream and 1.3 times for high bit-rate bitstream.

3.2.2.2   Image Processing Algorithm

[32] describes the process and results of mapping of two image processing algorithms, Wavelet encoding and decoding, and TIFF compression on

Kernel	No' of ops	II	IPC	Stages	Sched. Time (secs)
get_block1	27	1	27	10	44.5
get_block2	113	4	28.3	6	479.6
get_block3	119	4	29.8	6	459.4
get_block4	74	3	24.7	6	212.5
get_block5	95	3	31.7	7	243.0
get_block6	82	3	27.3	6	227.8
get_block7	97	3	32.3	7	234.2
get_block8	91	3	30.3	10	328.8
mc_chroma1	79	3	26.3	10	541.5
mc_chroma2	145	5	29	9	1783
Itrans1	24	1	24	13	60.6
Itrans2	63	2	31.5	11	204.8
alloc_picture1	33	3	11	2	48.0
alloc_picture2	45	5	9	2	194.7
avg_block	32	2	16	4	90.0
copy_mv	44	4	11	3	226.6

FIGURE 3.6: Scheduling results of kernels. (From B. Mei, F.J. Veredas, and B. Masschelein, "Mapping an h.264/avc decoder onto the ADRES reconfigurable architecture," In *Proceedings of International Conference on Field Programmable Logic and Applications,* © 2005 IEEE. With permission.)

Bitstream	Frames/s (VLIW)	Frames/s (ADRES)	Speed-up (kernel/overall)
foreman_bf_264k	13.5	25.4	4.2/1.88
foreman_nobf_308k	15.0	26.2	4.2/1.75
mobile_nobf_320k	15.9	30.9	4.3/1.94
mobile_nobf_1366k	9.47	13.3	4.3/1.41

FIGURE 3.7: Simulation results of the mapped H.264/AVC decoder (at 100MHz). (From B. Mei, F.J. Veredas, and B. Masschelein, "Mapping an h.264/avc decoder onto the ADRES reconfigurable architecture," In *Proceedings of International Conference on Field Programmable Logic and Applications,* © 2005 IEEE. With permission.)

ADRES in a systematic way. The experiments described in this work were carried out on different instances with different numbers of FUs, different word width, different setup for the local RFs, and different interconnects between the FUs. More specific details about those architectures are shown in Figure 3.8. Both applications were simulated on different ADRES architectures as well as on the DSP c64x with an optimized source code version for each architecture including the TI c64x. The results of those simulations are shown in Figure 3.9. In all cases except for one the ADRES architecture and its corresponding compiler DRESC performs better than the TI c64x. Only the 2x2 architecture is worse than the DSP for the Tiff2BW benchmark. This specific ADRES architecture has only half of the functional units of the TI c64x and the parallelism of the benchmark is fairly easy to exploit on the DSP c64x. For the wavelet transformation even the tradeoff of less FUs is compensated in a 2x2 architecture and the 2x2 ADRES architecture outperforms the TI c64x. Especially in the wavelet transformation a significant gain can be reached. For the 4x4 multimedia architecture the number of cycles was reduced to less than 25% compared to the TI c64x.

3.2.3 PACT-XPP

The PACT-XPP [11] is a high performance runtime coarse-grained reconfigurable processor. The XPP processor contains 64 ALU-PAEs (processing array elements) of 24 bit data within an 8x8 array. Each ALU-PAE contains three objects. The ALU-PAE performs several arithmetic, boolean, and shift operations. The Back-Register-object (BREG) provides routing channels for data and events from bottom to top, additional arithmetic functions and a barrel-shifter. For performance evaluation, a virtual platform including an ARM (ARM926EJ-S) micro-controller and a XPP processor coupled by an AMBA bus within MaxSim environment has been implemented in order to enable system verification [57]. The H.264 video decoder C-code for ARM is profiled and suitable functions are selected for implementation on the XPP processor. The remaining H.264 decoder functions that are in C run on the ARM microcontroller. This combined ARM+XPP solution ensures the implementation of a compliant H.264 video decoder. Four major functional blocks have been selected for this experiment. It contains Luma Reconstruction, Inverse Transition (Itrans) + Inverse Quantization (IQ), Color Conversion, and additionally upscaling for enlarging images. The selected functions are ported into XPP. Rests of H.264 functions remain in ARM. The selected functions are occupied over 50% of the total runtime cycles when it runs on ARM: Luma Reconstruction (40.16%), Itrans+IQ (13.2%), Color Conversion (4.37%). After they ported the modules into XPP, the result shows extremely improved performance: Luma Reconstruction (17.5 times), Itrans+IQ (47 times), and Color Conversion (13 times) respectively. Figure 3.10 depicts the performance comparison in execution cycles on 48x32 frame size. With the ported XPP

Template	Word Width	Number of FUs	Number of FUs with Load/Store capability	Number of RFs	Connections
4x4 multimedia	32	16	4	12	Nearest neighbor
4x4 wireless	64	16	4	13	mesh-plus routing
2x2 alternative	32	4	2	1 for 4words	MorphoSys routing
4x4_all alternative	32	16	4	12, connected to each FU	mesh-plus routing
4x4_4L alternative	32	16	4	4, each for 4 words	MorphoSys routing
4x4_16L alternative	32	16	4	4, each for 64 words	MorphoSys routing
8x8 alternative	32	64	8	16, each for 4 words	MorphoSys routing

FIGURE 3.8: Different ADRES architectures and their characteristics. (From M. Hartmann, V. Pantazis, T. Vander Aa, M. Berekovic, C. Hochberger, and B. de Sutter, "Still image processing on coarse-grained reconfigurable array architectures." In *Proceedings of IEEE/ACM/IFIP Workshop on Embedded Systems for Real-Time Multimedia,* © 2007 IEEE. With permission.)

Benchmark	Architecture	Cycles	Instructions	IPC	Utilization of the CGRA	Speedup to TI c64x
Tiff2BW	DSP TI c64x	4,547,947	26,529,081	6		1.00
	4x4 multimedia	2,273,967	24,729,217	10.9	68%	2.00
	4x4 wireless	3,410,949	26,718,897	7.8	49%	1.33
	2x2 alternative	8,811,517	24,444,802	2.8	70%	0.52
	4x4_all alternative	2,894,189	32,662,290	11.3	71%	1.57
	4x4_4L alternative	2,687,434	32,662,331	12.2	76%	1.69
	4x4_16L alternative	2,687,427	21,421,906	11.7	73%	1.69
	8x8 alternative	1,421,264	35,814,939	25.2	39%	3.20
Wavelet	DSP TI c64x	8,765,163	26,973,231	3.08		1.00
	4x4 multimedia	2,095,528	28,346,939	13.5	84%	4.18
	4x4 wireless	2,623,134	30,137,922	11.5	72%	3.34
	2x2 alternative	6,767,899	26,117,944	3.9	98%	1.30
	4x4_all alternative	3,091,148	35,451,612	11.5	72%	2.84
	4x4_4L alternative	2,885,585	35,022,239	12.1	76%	3.04
	4x4_16L alternative	2,792,529	34,202,225	12.2	76%	3.14
	8x8 alternative	1,627,106	46,342,177	28.5	45%	5.39

FIGURE 3.9: Comparison of the image processing benchmarks for different ADRES architectures and the TI c64x. (From M. Hartmann, V. Pantazis, T. Vander Aa, M. Berekovic, C. Hochberger, and B. de Sutter, "Still image processing on coarse-grained reconfigurable array architectures," In *Proceedings of IEEE/ACM/IFIP Workshop on Embedded Systems for Real-Time Multimedia,* © 2007 IEEE. With permission.)

H.264(48X32) Upscaled 10 frames	ARM	PACT	Speedup
Upscaling	98M	815.8K	120X
Luma Reconstruction	6M	342.2K	17.5X
Itrans + IQ	3M	64K	47X
Color Conversion	1M	77.5K	13X
Entire Decoder	123.5M	16.7M	7.4X

FIGURE 3.10: Speedup of H.264. (From M. Hartmann, V. Pantazis, T. Van-der Aa, M. Berekovic, C. Hochberger, and B. de Sutter, "Still image processing on coarse-grained reconfigurable array architectures," In *Proceedings of IEEE/ACM/IFIP Workshop on Embedded Systems for Real-Time Multimedia*, © 2007 IEEE. With permission.)

functions, the overall performance improved 7.4 times compared to the implementation with ARM.

3.2.4   PipeRench

PipeRench [74] is a coarse-grained reconfigurable fabric—an interconnected network of configurable logic and storage elements. PipeRench is designed to efficiently handle computations. Using a technique called pipeline reconfiguration, PipeRench improves compilation time, reconfiguration time, and forward compatibility. PipeRench's architectural parameters (including logic block granularity) optimize the performance of a suite of kernels, balancing the compiler's needs against the constraints of deep-submicron process technology. PipeRench is particularly suitable for stream-based media applications or any applications that rely on simple, regular computations on large sets of small data elements.

To evaluate PipeRench's performance [27], the authors chose nine kernels on the basis of their importance, recognition as industry performance benchmarks, and ability to fit into our computational model

- *Automatic target recognition (ATR)* implements the shape-sum kernel of the Sandia algorithm for automatic target recognition.

- *Cordic* implements the Honeywell timing benchmark for Cordic vector rotations.

- *DCT* is a 1D, 8-point discrete cosine transform.

- *DCT-2D* is a 2D discrete cosine transform.

- *FIR* is a finite-impulse response filter with 20 taps and 8-bit coefficients.

- *IDEA* implements a complete 8-round International Data Encryption Algorithm.

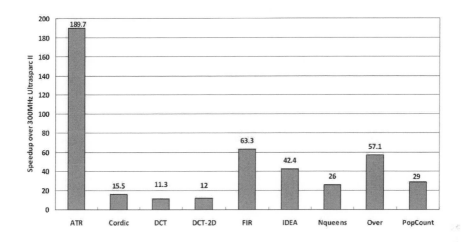

FIGURE 3.11: Speedup for nine kernels: raw speedup of a 128-bit fabric with 8-bit PEs and eight registers per PE. (From S.C. Goldstein, H. Schmit, M. Budiu, S. Cadambi, M. Moe, and R.R. Taylor, "PipeRench: a reconfigurable architecture and compiler," *IEEE Computer*, © 2000 IEEE. With permission.)

- *Nqueens* is an evaluator for the N queens problem on an 8x8 board.

- *Over* implements the Porter-Duff over operator.

- *PopCount* is a custom instruction implementing a population count instruction.

They compared 100-MHz PipeRench's performance with the performance of an Ultrasparc II running at 300 MHz. Figure 3.11 shows the raw speedup for each kernel. The results vary from 12 times to 189.7 times.

3.2.5 RDP

The Reconfigurable Data Path (RDP) [20] is composed of microprocessor cores with a high-performance coarse-grained reconfigurable data-path presented. The computational resources of the Reconfigurable Data Path (RDP) are able to implement complex arithmetic structures. An outline of the data-path is shown in Figure 3.12. The considered high-performance coarse-grained reconfigurable data-path and the respective mapping methodology have been introduced in [20]. The proposed coarse-grained RDP consists of a set of word-level hardwired Coarse-Grained Components (CGC) units, a reconfigurable interconnection network, a register bank, configuration RAMs (CR), Address Generation Units (AGUs) and a control unit. The data-width of the RDP is typically 16-bits, although different word-level bit-widths are supported.

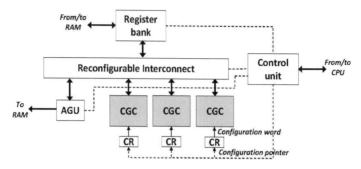

FIGURE 3.12: Outline of a CGC-based data-path. (From G. Dimitroulakos, N. Kostaras, M. Galanis, and C. Goutis, "Compiler assisted architectural exploration for coarse grained reconfigurable arrays," In *Proceedings of Great Lakes Symposium on VLSI*, © 2007 IEEE. With permission.)

For the performance evaluation of RDP [24], the eight real-world DSP applications, described in C language, used in the experiments are given in Figure 3.13. A brief description of each application is given in the second column, while in the third one the input sequence used is presented. The execution times and the overall application speedups for the eight applications are presented in Figure 3.14. $Time_{sw}$ represents the software execution time of the whole application on a specific microprocessor (*Proc.*). The ideal speedup (*Idealsp.*) is the application speedup that would ideally be achieved, according to Amdahl's Law, if application's kernels were executed on the RDP in zero time. $Time_{system}$ corresponds to the execution time of the application in-

Application	Description	Input
JPEG enc.	Still-image JPEG encoder	256x256 byte image
OFDM trans.	IEEE 802.11a OFDM transmitter	4 payload symbol
Compressor	Wavelet-based image compressor	512x512 byte image
Cavity det.	Medical imaging technique	640x400 byte image
Edge det.	Edge detection in images	128x128 byte image
JPEG dec.	Still-image JPEG Decoder	227x149 byte image
GSM enc.	Speech encoder	clinton.pcm
GSM dec.	Speech decoder	clinton.pcm.run.gsm

FIGURE 3.13: Applications' characteristics. (From M. Galanis, G. Dimitroulakos, and C. Goutis, "Speedups and energy savings of microprocessor platforms with a coarse-grained reconfigurable datapath," In *Proceedings of International Parallel and Distributed Processing Symposium*, © 2007, IEEE. With permission.)

cluding the communication overhead between the processor and the RDP. All execution times are normalized to the software execution times on the ARM7. The *Sp.* is the estimated application speedup ($Time_{sw}$ / $Time_{system}$). From the results given in Figure 3.14, it is evident that significant overall performance improvements are achieved when critical software parts are mapped on the CGCs. These speedups range from 1.74 to 3.94. It is noticed from Figure 3.14 that the largest overall application performance gains are achieved for the ARM7-based architectures as the ARM7 has the lowest clock frequency and exhibits the highest Cycles Per Instruction (CPI) among the three ARM-based systems. The average application speedup of the eight DSP benchmarks for the ARM7 systems is 2.90, for the ARM9 is 2.68, while for the ARM10 systems is 2.53. Thus, even when the CGC based data-path is coupled with a modern embedded processor, as the ARM10, which is clocked at a higher clock frequency, the application speedup over the execution on the processor core is significant. The average overall application speedup for all the microprocessor systems is 2.70.

3.3 Summary

In embedded system-on-chip, different kinds of IP-types show the trade-off between performance and flexibility. However, the incompatible demands (high performance and flexibility) can be compromised by CGRA because CGRAs are implemented by domain-specific design method. In this chapter, we present performance evaluation examples of some CGRAs (MorphoSys, ADRES, PACT-XPP, PipeRench and RDP) for showing quantitative comparison among CGA, FPGA, GPP and ASIC.

Application	Processor	Time$_{sw}$	Ideal Speedup	Processor/RDP	
				Time$_{system}$	Speedup
JPEG enc.	ARM7	1.000	3.96	0.272	3.68
	ARM9	0.461	3.24	0.148	2.93
	ARM10	0.301	3.16	0.092	2.67
OFDM trans.	ARM7	1.000	3.54	0.305	3.28
	ARM9	0.485	3.43	0.146	3.13
	ARM10	0.344	3.23	0.089	3.13
Compressor	ARM7	1.000	2.51	0.450	2.22
	ARM9	0.424	2.32	0.197	2.02
	ARM10	0.283	2.21	0.132	1.75
Cavity det.	ARM7	1.000	2.38	0.493	2.03
	ARM9	0.480	2.29	0.241	1.87
	ARM10	0.355	2.17	0.166	1.74
Edge det.	ARM7	1.000	2.61	0.407	2.46
	ARM9	0.498	2.54	0.198	2.36
	ARM10	0.367	2.49	0.132	2.27
JPEG dec.	ARM7	1.000	4.17	0.254	3.94
	ARM9	0.418	3.85	0.108	3.64
	ARM10	0.273	3.64	0.067	3.32
Gsm enc.	ARM7	1.000	3.05	0.348	2.87
	ARM9	0.426	2.93	0.145	2.77
	ARM10	0.295	2.88	0.090	2.67
Gsm dec.	ARM7	1.000	2.82	0.364	2.75
	ARM9	0.422	2.77	0.146	2.71
	ARM10	0.292	2.74	0.089	2.67
Average			2.96		2.70
Geo. mean			2.90		2.64

FIGURE 3.14: Execution times and application speedups. (From M. Galanis, G. Dimitroulakos, and C. Goutis, "Speedups and energy savings of microprocessor platforms with a coarse-grained reconfigurable datapath," In *Proceedings of International Parallel and Distributed Processing Symposium*, © 2007, IEEE. With permission.)

Chapter 4

Base CGRA Implementation

4.1 Introduction

We have first designed a conventional CGRA as the base architecture and implemented it at the RT-level. This conventional architecture will be used throughout this book as a reference for quantitative comparison with our cost-effective approaches.

4.2 Reconfigurable Array Architecture Coupling with Processor

Many kinds of coarse-grained reconfigurable architecture have been proposed with the increasing interests in reconfigurable computing until 2001 [30]. These CGRAs consist of a microprocessor, a Reconfigurable Array Architecture (RAA), and their interface. We can consider three ways of connecting the RAA to the processor [10]. First, the array can be connected to a bus as an 'Attached IP' [11,19,35,46,70,75] shown in Figure 4.1(a). Secondly, the array can be placed next to the processor as a 'Coprocessor' [12,25,61] as shown in Figure 4.1(b). In this case, the communication is done using a protocol similar to those used for floating point coprocessors. Finally, the array can be placed inside the processor like a 'FU (Functional Unit)' [4,7,9,24,26] as shown in Figure 4.1(c). In this case, the instruction decoder issues special instructions to perform specific functions on the reconfigurable array as if it were one of the standard functional units of the processor.

We have implemented the first type of reconfigurable architecture connecting the RAA as an Attached IP as shown in Figure 4.1(a). In this case, the speed improvement using the RAA may have to compensate for significant communication overhead. However, the main benefit of this type is the ease of constructing such a system using a standard processor and standard reconfigurable array without any modification. It consists of a RISC processor, a main memory block, a DMA controller, and an RAA. The RISC processor is a 32-bit processor which is small and simple with three pipeline stages and the

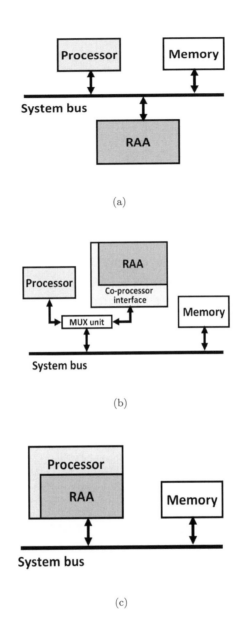

FIGURE 4.1: Basic types of reconfigurable array coupling: (a) attached IP; b) coprocessor; (c) functional unit.

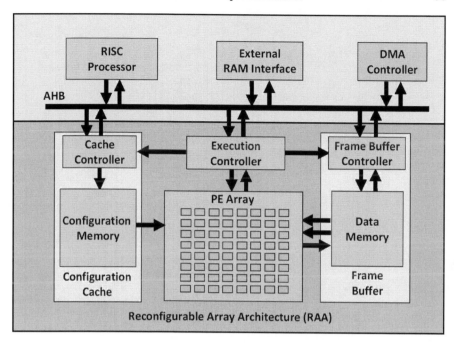

FIGURE 4.2: Block diagram of base CGRA.

communication bus is AMBA AHB [16], which couples the RISC processor and the DMA controller as master devices and the RAA as a slave device. The RISC processor executes control intensive, irregular code segments and the RAA executes data-intensive kernel code segments. The block diagram of the entire reconfigurable architecture is shown in Figure 4.2.

4.3 Base Reconfigurable Array Architecture

Base RAA is similar to MorphoSys [75], which is a very representative CGRA showing high performance and flexibility as well as physical implementation. The difference from MorphoSys is that the proposed architecture supports both SIMD and MIMD execution models whereas the memory structure (frame buffer and configuration cache) of MorphoSys supports only the SIMD model. The SIMD model is efficient for data parallelism since it saves configurations and cache storage by sharing an instruction for multiple data. But its execution models are limited in that each individual PE cannot execute different instructions independently at the same time. Therefore, we take MIMD-style CGRA in which each PE can be configured separately to

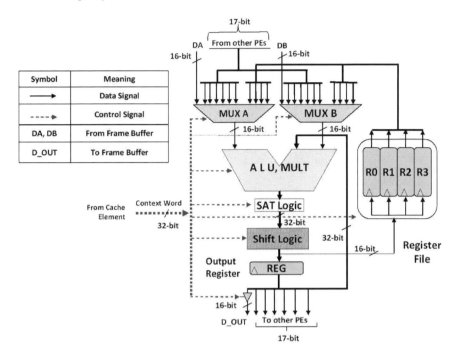

FIGURE 4.3: Processing element structure of base RAA.

facilitate processing its own instructions. Since it allows more versatile configurations than their SIMD-style siblings, we adopt more general forms of loop pipelining [51] through simultaneous execution of multiple iterations of a loop in a pipeline.

Base architecture specification is determined by our target application domain including audio/video codec as well as various benchmark kernels. Detailed features of each component of the architecture are as follows.

4.3.1   Processing Element

Each PE is a dynamically reconfigurable unit executing arithmetic and logical operations. The inner structure of a PE is shown in Figure 4.3. A PE contains a 16-bit ALU, 16 x 16-bit array multiplier, shift logic, Arithmetic saturation (SAT_Logic), multiplexors and registers.

4.3.2   PE Array

The PE array is an 8x8 reconfigurable array of PEs, which we think is big enough for most of the applications considered in our experiments. We assume that computation model of the array is loop pipelining based on temporal map-

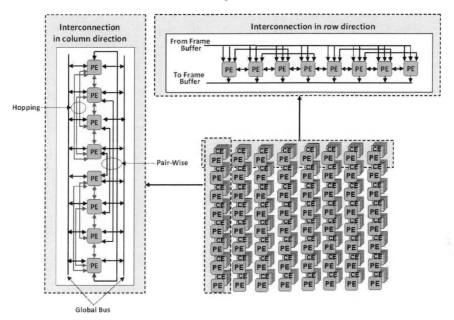

FIGURE 4.4: Interconnection structure of PE array.

ping [51] for high performance—each iteration of application kernel (critical loop) is mapped onto each column of square array. Therefore, in this PE array, columns have more interconnection than rows. Figure 4.4 shows interconnection structure of the PE array. The interconnection in rows is used mainly for the communication taking care of loop-carried dependencies. Columns and rows have nearest-neighbor and hopping interconnections for connectivity between two PEs in a half column and a half row. In addition, each column has pair-wise interconnections and two global buses for connectivity between two half columns. Each row shares two read-buses and one write-bus.

4.3.3 Frame Buffer

Frame buffer (FB) of MorphoSys does not support concurrency between the load of two operands and the store of result in a same column, since it is not needed in SIMD-style mapping. However, in the case of MIMD-style execution, concurrent load and store operations can happen between different loop iterations. So our FB has two sets of buffers, each having three banks: one bank connected to the write bus and the other two banks connected to the read buses. However, any combination of one-to-one mapping between the three banks and the three buses is possible.

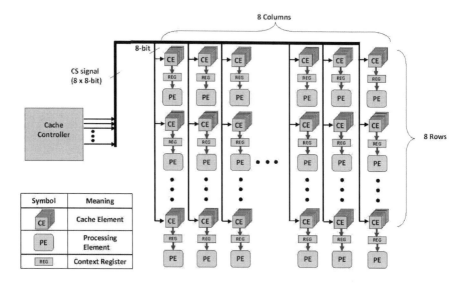

FIGURE 4.5: Distributed configuration cache structure.

4.3.4   Configuration Cache

The context memory of MorphoSys is designed for broadcasting configuration. So PEs in the same row or column share the same context word for SIMD-style operation [75]. However, in the case of MIMD-style operation, each PE can be configured by different context word. Our configuration cache is composed of 64 Cache Elements (CEs) and a cache controller for controlling the CEs (Figure 4.5). Each CE has 32 layers, each of which stores a context that configures the corresponding PE. The context register between a PE and a CE is used to keep the cache access path from being the critical path of the CGRA.

4.3.5   Execution Controller

Controlling the PE array execution directly from the main processor through AMBA AHB will cause high overhead in the main processor. In addition, the latency of the control will degrade the performance of the whole system, especially when dynamic reconfiguration is used. So a separate control unit is necessary to control the execution of the PE array every cycle. The execution controller receives the encoded control data from the main processor. The control data contains read/write mode and addresses of frame buffer and cache for guaranteeing correct operations of the PE array.

4.4 Breakdown of Area and Delay Cost

We have implemented the base architecture shown in Figure 4.1 at the RT-level with VHDL. We have synthesized a gate-level circuit from the VHDL description and analyzed area and delay cost. The synthesis has been done using Design Compiler [18] with 0.18 μm technology. We have used Design-Ware [18] library for the multipliers (carry-save array synthesis model) and SRAM Macro Cell library is used for the frame buffer and configuration cache.

4.4.1 Area and Delay

As shown in Figure 4.6(a), the RAA occupies as much as 90 % of the total area of the CGRA. Figure 4.6(b) shows more detailed area breakdown in the RAA. The PE array occupies as much as 70.5 % of the total area of the RAA, which is mainly due to heavy computational resources such as ALU, multiplier, etc. in each PE. The critical path of the entire RAA is also in the PEs and its delay is given by

$$T_{critical\_path} = T_{Multiplexor} + T_{Multiplier} + T_{Shift\_logic} + T_{others} \qquad (4.1)$$

$$(8.96\text{ns} = 0.32\text{ns} + 5.21\text{ns} + 1.42\text{ns} + 1.78\text{ns})$$

From the area and delay cost breakdown of the RAA as shown in Figure 4.6 and Figure 4.7, we see that PE array design is crucial for cost-effective design. In the case of area, Figure 4.7(a) shows that multiplier occupies about 33.4% of the total area in a PE. In the case of delay, the multiplier again takes as much as 58.12 % (Figure 4.7(b)). Therefore, in our PE design, the multiplier is considered to be area-critical and delay-critical resource.

4.5 Summary

In this chapter, we describe the base CGRA structure. First of all, coupling structure between RAA and Processor is presented in the aspect of entire system. Then computing elements (PE) and memory elements (Frame Buffer and Configuration Cache) of RAA are illustrated to show the mechanism of CGRA operation. In addition, we present RT-level implementation of CGRA and their area/delay analysis for quantitative evaluation.

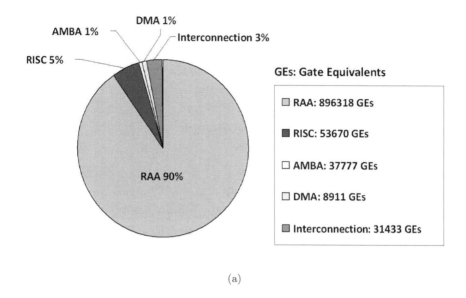

(a)

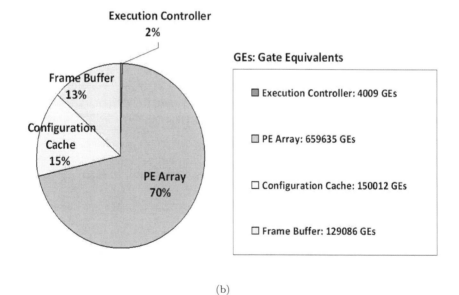

(b)

FIGURE 4.6: Area cost breakdown for CGRA: (a) entire CGRA; (b) RAA.

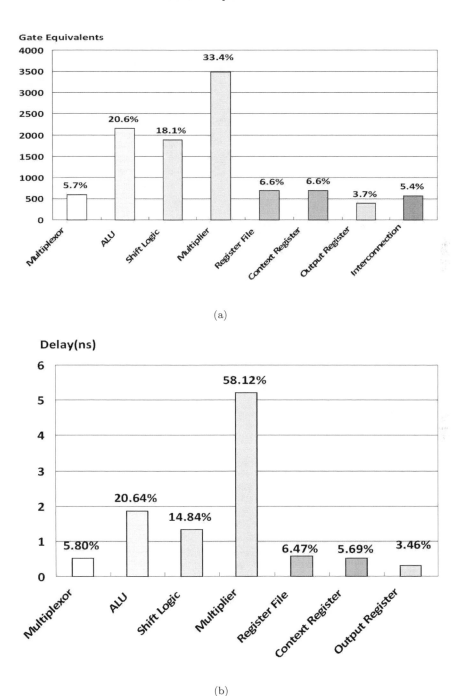

FIGURE 4.7: Cost analysis for a PE: (a) area; (b) delay.

Chapter 5

Power Consumption in CGRA

5.1 Introduction

Coarse-grained reconfigurable architectures are considered appropriate for embedded systems because they can satisfy both flexibility and high performance requirements. However, power consumption is also crucial for the reconfigurable architecture to be used as a competitive processing core in embedded systems. This chapter shows breakdown of power consumption in CGRA for two cases—the base CGRA mentioned in Chapter 4 and ADRES [60]. Based on the power analysis results, we describe which component of CGRA consumes dominant power and why power consumption has been a serious concern for reliability of CGRA-based embedded systems.

5.2 Breakdown of Power Consumption in CGRA

5.2.1 Base CGRA

As described in Chapter 4, we have implemented the base architecture at the RT-level with VHDL. We have synthesized a gate-level circuit from the VHDL description and analyzed power cost. The synthesis has been done using Design Compiler [18] with 0.18 μm technology. ModelSim [17] and PrimePower [18] have been used for gate-level simulation and power estimation. To analyze power consumption of base CGRA described in Chapter 4, we have used 2D-FDCT as the kernel for simulation-based power measurement. The simulation has been done under the typical operating condition of 100 MHz frequency, 1.8 V Vdd, and 27°C temperature.

As can be observed from Figure 5.1(a), the RAA spends about 92.09% of the total power consumed in CGRA. Figure 5.1(b) shows more detailed power breakdown in the RAA. The RAA spends about 50.8% of its total power in the PE array, which consists of many components such as ALUs, multipliers, shifters and register files. The PE array consumes most of the power, which is natural because coarse-grained architecture aims to achieve high performance and flexibility with plenty of resources. The configuration cache spends about 45.3% of the overall power, which is the second largest. Even though the frame buffer uses the same kind of SRAM as the configuration cache, it consumes much less power (3.4%). This is because the configuration cache performs read operations frequently to load the context words, one for each PE, whereas the frame buffer performs load/store operations less frequently to access data on row basis rather than for every PE.

5.2.2   ADRES

The power figure of the ADRES instance is adressed in [21]. The 4x4 ADRES instance XML description is transformed into VHDL files and is synthesized in the Synthesize part. Synopsys Physical Compiler [18] is used to create the netlist for 90nm CMOS library. ModelSimRTL simulator [17] is used to simulate the generated RTL VHDL files and to obtain the switching activity figures needed for RTL power calculations. This switching activity is what an RTL HDL simulator also generates for the power analysis tool. The switching activities obtained after simulations are annotated on the gate level design created in the synthesize part. The toggling file and the gate level design are used by the PrimePower [18] to estimate power.

The detailed power breakdown for IDCT are depicted in Figure 5.2. The components with _vliw postfix are located in the VLIW section and are operational in both VLIW and CGA mode, while those with the _cga postfix are located in the CGA section. The power chart in Figure 5.2 shows that the Configuration Memories (CMs), Funtional Units (FUs) and Data Register Files (DRFs) consume most of the power. The VLIW control unit (cu vliw) and Predicate Register Files (PRFs) consume the least amount of power as these are the smallest components.

The power measurements of ADRES were based on the core architecture only. To get a power prediction of the complete ADRES processor including instruction cache and data memories we simulated the activities of these blocks. For the IDCT the data memories were activated for 100% of the cycles and the instruction cache for 4.3% respectively. According to the data sheets of SRAM memory models this results in a power consumption of 0.03mW/MHz for the data memories and 0.05 mW/MHz for the instruction cache. The power consumption of the not activated memory modules is negligible and assumed as zero. The distribution of all the components in the processor for the IDCT benchmark is shown in Figure 5.3. It shows that in addition to the FUs, DRFs and CMs, the data memory (DMEM) consumes almost 10% of power. The I-cache power consumption is negligible.

Based on such a power analysis, we can see that the analysis result is similar to the case of base CGRA mentioned in the previous subsection. The entire ADRES spends about 40% of its total power consumed in the CGA like the PE array of the base CGRA. In addition, the CMs spend about 33.44% of the overall power, which is the second largest like configuration cache of the

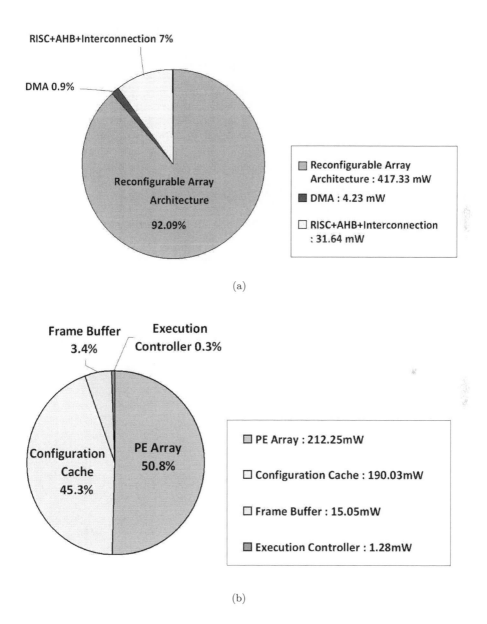

FIGURE 5.1: Power cost breakdown for CGRA running 2D-FDCT: (a) entire CGRA; (b) RAA.

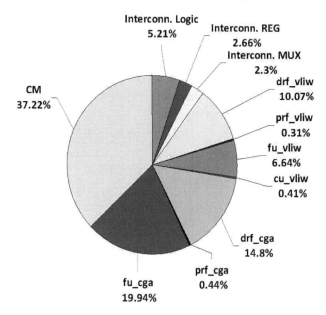

FIGURE 5.2: Power cost breakdown for only ADRES core with IDCT. (From A. Kanstein, F. Bouwens, M. Berekovic, and G. Gaydadjiev, "Architectural exploration of the ADRES coarse-grained reconfigurable array," In *Lecture Notes in Computer Science*, pages 1–13, © 2007. With kind permission of Springer Science+Business Media.)

base CGRA since frequent reconfiguration of the CGA causes many CM-read operations and in turn causes much power consumption.

5.3   Necessity of Power-Conscious CGRA Design

The previous section shows that the reconfigurable array part and the configuration cache (memory) spend most of the total power. This is due to the fact that CGRA is composed of many computational resources and storage elements. In the case of the reconfigurable array, it plays an important role for high performance and flexibility with plenty of resources but it suffers from large power consumption. Furthermore, if array size increases with including more functional resources for enhancing performance and flexibility, it is explicit that the power consumption by the array also increases. It means that the reconfigurable array design is very important in the trade-off among performance, flexibility, and power.

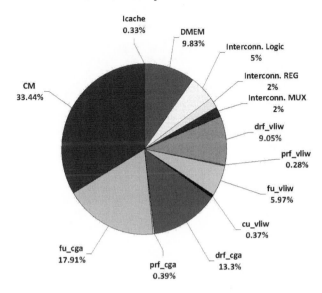

FIGURE 5.3: Overview of entire ADRES processor power consumption with IDCT. (From A. Kanstein, F. Bouwens, M. Berekovic, and G. Gaydadjiev, "Architectural exploration of the ADRES coarse-grained reconfigurable array," In *Lecture Notes in Computer Science*, pages 1–13, © 2007. With kind permission of Springer Science+Business Media.)

In the case of configuration cache, it is the main component in CGRA that provides distinct feature for dynamic configuration. However, such dynamic reconfiguration of CGRA causes many SRAM-read operations in configuration cache as shown in Figure 5.1 and 5.3. In addition, in [74], the authors have fabricated a CGRA (PipeRench) in a 0.18 μm process. Their experimental results show that the power consumption is significantly high due to the dynamic reconfiguration requiring frequent configuration memory access. It means that power consumption by configuration cache (memory) is serious overhead compared to other types of IP cores such as ASIC or ASIP. Therefore, reducing power consumption in the configuration cache has been a serious concern for reliability of embedded systems. From the next chapter, we address the problems of reducing power in the reconfigurable array and configuration cache based on architecture optimization.

5.4 Summary

This chapter shows breakdown of power consumption in the base CGRA (Chapter 4) and ADRES [60]. The power analysis shows that PE array/FU array and configuration cache/memory consumes dominant power. Based on the results, we can understand why power consumption has been a serious concern for reliability of CGRA-based embedded systems.

Chapter 6

Low-Power Reconfiguration Technique

6.1 Introduction

Coarse-grained reconfigurable architectures are considered appropriate for embedded systems because they can satisfy both flexibility and high performance requirements. However, power consumption is also crucial for the reconfigurable architecture to be used as a competitive processing core in embedded systems. In this chapter, we describe an efficient power-conscious architectural technique called reusable context pipelining (RCP) [43] to reduce power consumption in configuration cache. RCP is a universal approach in reducing power and enhancing performance for CGRA because it can be achieved by closing the power-performance gap between low-power-oriented spatial mapping and high performance-oriented temporal mapping. In addition, hybrid configuration cache structure is more efficient than previous one in terms of memory size.

6.2 Motivation

In this section, we present the motivation of our power-conscious approaches. The main motivation is due to the characteristics of loop pipelining (spatial mapping and temporal mapping) [51] based on MIMD-style execution model.

6.2.1 Loop Pipelining

To represent the characteristics of loop pipelining [51], we examine the difference between SIMD and MIMD in the RAA with a simple example. We assume a mesh-based 4x4 coarse-grained reconfigurable array of PEs, where a PE is a basic reconfigurable element composed of an ALU, an array multiplier, etc. and the configuration is controlled by the words stored in the CE as shown in Figure 6.1(a). In addition, we assume that Frame Buffer has simply one set having three banks and two read-ports and one write-port, supporting any combination of one-to-one mapping between the three banks and the three buses. Figure 6.1(b) shows such a Frame Buffer and data bus structure, where the PEs in each row of the array share two read buses and one write bus. The 4x4 array has nearest neighbor interconnections as shown in Figure 6.1(c) and each row or each column has a global bus as shown in Figure 6.1(d).

Consider a square matrix X and Y, both of order N, and the computation of Z, an N element vector, given by

$$Z(i) = K \times \sum_{j=0}^{N-1} \{(X(i,j) + Y(i,j)) \times C(j)\} \tag{6.1}$$

where $i, j = 0,1,,N\text{-}1$, $C(j)$ is a constant vector, and K is a constant. Consider $N = 4$ for the mapping of the computation defined in Eq. 6.1 on our 4x4 PE array and let the computation be given as a C-program (Figure 6.2(a)). It is assumed that the input matrix X, Y, constant vector C and output vector Z are stored in the arrays $x[i][j]$, $y[i][j]$, $c[j]$ and $z[i]$, and $z[i]$ is initialized to zero. Figure 6.2(b) shows parallelized code for execution on the array as shown in Figure 6.3, where we assume that matrix X and Y have been loaded into the Frame Buffer (FB) and all of the constants (C and K) have been already saved in a register file of each PE. Vector Z is stored in the FB after it has been processed by the PE array as shown in Figure 6.3(a).

The SIMD-based scheduling enables parallel execution of multiple loop iterations as shown in Figure 6.3(c), whereas the MIMD-based scheduling enables loop pipelining as shown in Figure 6.3(d). The first row of Figure 6.3(c) represents the direction of configuration broadcast. The second row of Figure 6.3(c) and the first row of Figure 6.3(d) indicate the schedule time in cycles from the start of the loop. In the case of SIMD model, load and addition operations in PEs are executed on all columns till 4th cycle with broadcast in column direction. Then the PEs in a row perform the same operation with broadcast in row direction. In the case of loop pipelining, PEs in the first column perform load and addition operations in the first cycle and then perform multiplications in the second cycle. In the next two cycles, the PEs in the first column perform summations, while the PEs in the next column perform multiplication and summation operations. When the first column performs the multiplication/store operation in the 5th cycle, the fourth column performs multiplication. Comparing the latency, SIMD takes three more cycles.

As shown in this example, SIMD model does not utilize PEs efficiently since all data should be loaded before the computations of the same type are performed synchronously. On the other hand, since MIMD allows any type of computations at any moment, it does not need to wait for a specific data to be loaded but can process other data that is readily available. Loop pipelining is an effective way of exploiting this fact, thereby utilizing PEs better. The loop pipelining in the example of Figure 6.3 improves the performance by

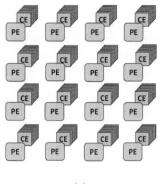

(a)

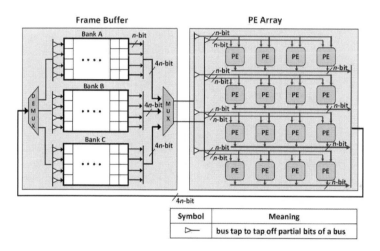

(b)

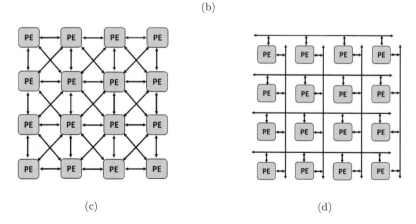

(c)   (d)

FIGURE 6.1: A 4x4 reconfigurable array: (a) distributed cache structure; (b) frame buffer and data bus; (c) nearest neighbor interconnection; and (d) global bus interconnection.

```
for (i = 0; i <= 3; i = i+1)
{
    for (j = 0; j <= 3; j = j+1)
        z[i] = (x[i][j]+y[i][j])*c[j] + z[i];
    z[i] = K* z[i];
}
```

(a)

```
for (i = 0; i <= 3; i = i+1)
{
    t1 = x[i][0]+y[i][0] ;
    t2 = x[i][1]+y[i][1] ;      LD/+
    t3 = x[i][2]+y[i][2] ;
    t4 = x[i][3]+y[i][3] ;

    t1 = t1*c[0];
    t2 = t2*c[2];               ×
    t3 = t3*c[3];
    t4 = t4*c[4];

    tmp1 = t1+ t2 ;             1+
    tmp2 = t3+ t4 ;

    z[i] = tmp1+ tmp2 ;         2+

    z[i] = K*z[i]               ×/ST
}
```

(b)

FIGURE 6.2: C-code of Eq. 6.1: (a) before parallelization and (b) after parallelization.

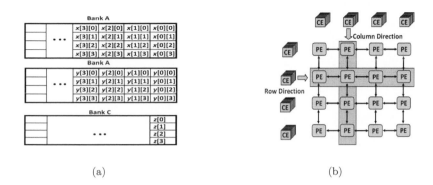

(a)   (b)

Broadcast	Column Direction				Row Direction			Column Direction			
Cycle Time	1	2	3	4	5	6	7	8	9	10	11
Column#1	LD/+	NOP	NOP	NOP	×	1+	2+	×/ST	NOP	NOP	NOP
Column#2		LD/+	NOP	NOP	×	1+	2+	NOP	×/ST	NOP	NOP
Column#3			LD/+	NOP	×	1+	2+	NOP	NOP	×/ST	NOP
Column#4				LD/+	×	1+	2+	NOP	NOP	NOP	×/ST

(c)

Cycle Time	1	2	3	4	5	6	7	8
Column#1	LD/+	×	1+	2+	×/ST	NOP	NOP	NOP
Column#2		LD/+	×	1+	2+	×/ST	NOP	NOP
Column#3			LD/+	×	1+	2+	×/ST	NOP
Column#4				LD/+	×	1+	2+	×/ST

Symbol	Meaning
LD/+	Data Load and Addition
NOP	No Operation
×	Multiplication
1+, 2+	Addition
×/ST	Multiplication and Store

(d)   (e)

FIGURE 6.3: Execution model for CGRA: (a) operand and result data in FB; (b) configuration broadcast; (c) SIMD model; (d) loop pipelining schedule; and (e) symbol definition.

three cycles compared to the SIMD, but for loops with more frequent memory operations, it will have higher performance improvement.

6.2.2   Spatial Mapping and Temporal Mapping

When mapping kernels onto the reconfigurable architecture with loop pipelining, we can consider two mapping techniques [51]: spatial mapping and temporal mapping. Figure 6.4 shows the difference between the two techniques with the previous example. In the case of temporal mapping (Figure 6.4(a)), like the previous illustration of loop pipelining in Figure 6.3(d), a PE executes multiple operations within a loop by changing the configuration dynamically. Therefore, complex loops having many operations with heavy data dependencies can be mapped better in temporal fashion, provided that the configuration cache has sufficient layers to execute the whole loop body.

In the case of spatial mapping, a loop body is spatially mapped onto the reconfigurable array implying that each PE executes a fixed operation with static configuration as shown in Figure 6.4(b). The advantage of spatial mapping is that it may not need reconfiguration during execution of a loop. As can be seen from Figure 6.4, spatial mapping needs only one or two cache

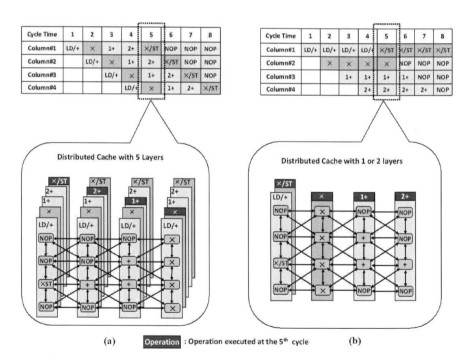

FIGURE 6.4: Comparison between temporal mapping and spatial mapping: (a) temporal mapping and (b) spatial mapping.

layers whereas temporal mapping needs 4 cache layers. One disadvantage of spatial mapping is that spreading all the operations of the loop body over the limited reconfigurable array may require too many resources. Moreover, data dependencies between the operations should be taken care of by allocating interconnect resources to provide a path and inserting registers (or using PEs) in the path to synchronize the arrival of operands. Therefore, if the loop is simple enough to map the loop body to the limited reconfigurable array and there is not much data dependency between the operations, then spatial mapping is the right choice. The effectiveness of the mapping strategies depends on the characteristics of the target architecture as well as the target application.

6.3 Individual Approaches to Reduce Power in Configuration Cache

In this section, we suggest individual power-conscious approaches for two different execution models (spatial mapping and temporal mapping) and describe their limitations. These approaches achieve the goal by making use of the characteristics of spatial mapping and temporal mapping [47, 67, 68].

6.3.1 Spatial Mapping with Context Reuse

Because most power consumption in the configuration cache is due to memory read-operations, one of the most effective ways to achieve power reduction in the configuration cache is to reduce the frequency of read operations.

Even though temporal mapping is more efficient in mapping complex loops onto the reconfigurable array, it requires many configuration data layers for each PE and performs power consuming read-operations in every cycle. On the other hand, spatial mapping does not need to read a new context word from the cache every cycle because each PE executes a fixed operation within a loop. As shown in Figure 6.5, if a context register between a CE and a PE is implemented by a gated clock, one spatial cache[1] read-operation is enough in spatial mapping to configure PEs for static operations with fixed output of the context register caused by non-oscillated clock. In summary, spatial mapping with context reuse is more efficient than temporal mapping from the viewpoint of power consumption in configuration cache. However, all kinds of loops cannot be spatially mapped because of the limitation of the spatial mapping. Moreover, if we consider performance alone, temporal mapping is

---

[1]We use the term 'spatial cache'. Spatial cache is connected to context registers implemented by gated clock. The term 'spatial' means that such configuration cache is used for spatial mapping with context reuse. This naming is to differentiate spatial cache from general configuration cache.

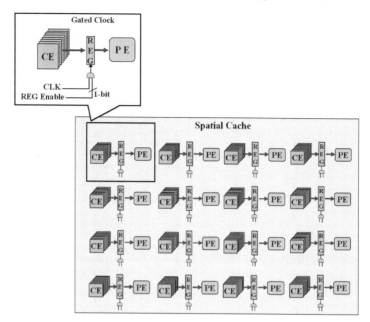

FIGURE 6.5: Configuration cache structure for context reuse.

a better choice for loops having long and complex loop body. In the next subsection, we propose a new cache structure and mapping technique that reduce power consumption while retaining the merits of temporal mapping.

6.3.2   Temporal Mapping with Context Pipelining

As shown in Figure 6.4(a), in temporal mapping with loop pipelining, operations flow column by column from left to right. In Figure 6.4(a), for example, the first column executes 'LD/+' in the first cycle and then in the second cycle, the second column executes 'LD/+' while the first column executes '×'. In temporal mapping, there is no need for a PE to have a CE. Instead, only PEs in the first column have CEs and the context word can be fetched from the left neighboring column. By organizing a pipelined cache structure as shown in Figure 6.6, we can propagate the context words column by column through the pipeline. In this way, we can remove most of the CEs from the array keeping temporal cache[2], thereby saving power consumption without any performance degradation. In summary, temporal mapping with context pipelining can efficiently support long and complex loops reducing power consumption in

---

[2]We use the term 'temporal cache'. Temporal cache is composed of the cache elements connected to the PEs in the first column. The term 'temporal' means that such CEs are used for temporal mapping with context pipelining. This naming is to differentiate temporal cache from general configuration cache and spatial cache.

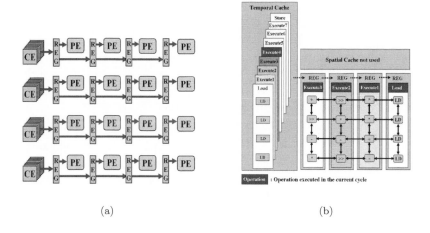

(a)   (b)

FIGURE 6.6: Cache structure for context pipelining: (a) cache structure and (b) context pipelining.

configuration cache. However, temporal mapping with context pipelining still needs cache-read operations for providing context words to the first column of PE array whereas spatial mapping with context reuse can remove cache-read operation after initial cache-read operation.

6.3.3   Limitation of Individual Approaches

As mentioned in the previous section, even though individual low-power techniques provide solutions to reduce power consumption for spatial mapping and temporal mapping, each case has both advantages and disadvantages. Spatial mapping with context reuse only need one cache-read operation for initialization but it cannot support the complex loops that cannot be spatially mapped. However, temporal mapping with context pipelining supports such complex loops but cache-read operations still remain in context pipelining for the running time. Therefore we should consider the trade-off between performance and power while deploying these techniques. We can consider two ways to close the gap between spatial mapping and temporal mapping. One is to implement more complex architecture to support high performance spatial mapping by adding additional interconnections or global register files for data dependency. However, in this case the area cost and mapping complexities will increase. Another way is to implement low-power temporal mapping taking advantage of spatial mapping with negligible over-head. However, the problem is how to implement this method. In the next section, we propose a new technique to guarantee the advantage of spatial mapping and temporal mapping. This is achieved by merging the concept of context reuse into context pipelining.

6.4 Integrated Approach to Reduce Power in Configuration Cache

Filling the gap between two mappings means that context pipelining is executed by reusable context words. However, it means conjunction of two mappings that are contrary to each other. This is because spatial mapping with context reuse requires spatially static position of each context whereas temporal mapping with context pipelining is performed with temporally changed context words. To solve this contradiction, we propose to add circular interconnection between the first PE and the last PE in the same row and suggest a reusable context pipelining using this interconnection.

6.4.1 Reusable Context Pipelining

Reusable context pipelining (RCP) means that reusable context words in spatial cache are pipelined through context registers as context pipelining. Figure 6.7(a) depicts the proposed configuration cache structure for RCP. Even though it is similar to the structure of Figure 6.5 (spatial mapping with context reuse), the new one has two context registers ('R1' and 'R2') connected to each PE, circular interconnections and less cache layers whereas the original model had one context register and more cache layers.

The circular interconnections and the context registers are necessary for pipelining of reusable context words from spatial cache. Figure 6.7(b) shows the detailed structure between a CE and a PE for RCP. A multiplexer ('MUX') is added between context registers ('REG1' and 'REG2') and PE for selecting one of the context registers or 'Zero'. Each context register is connected to each multiplexer ('MUX 1' or 'MUX 2') having two inputs: context word from left context register and context word from spatial cache. The input from spatial cache is for loading a reusable context word to the context register and the input from left context register is for pipelining execution of the loaded context word in left context register. Each select signal ('Select #1' or 'Select #2') connects one from two inputs to the single output connected with right context registers. Each context register is implemented by gated clock for holding the output as well as reducing the wasteful power consumption. All of the select-signals of the multiplexers are generated by cache control unit.

To present the detailed process of RCP, it is assumed that the matrix-vector multiplication given as Eq. 6.1 is mapped onto the proposed structure like the one in Figure 6.8. Figure 6.8(a) shows the context words stored in

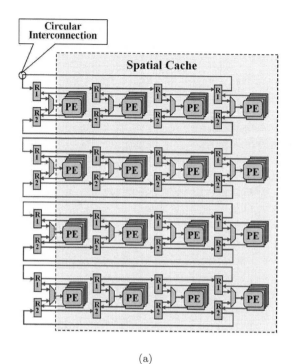

(a)

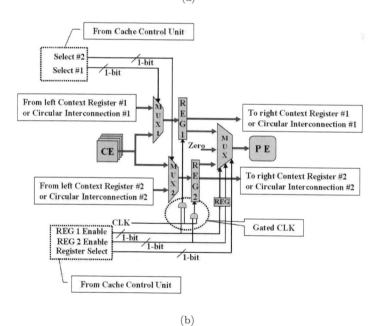

(b)

FIGURE 6.7: Proposed configuration cache structure: (a) entire structure and (b) connection between a CE and a PE.

spatial cache for RCP and Figure 6.8(b)–(i) shows the RCP process from the first cycle to the eighth cycle. Before starting execution, the context words of first layer in spatial cache are loaded into the first context registers ('REG 1'). At the first cycle, the PEs in the first column perform 'Load' and the context word ('Store') in spatial cache is loaded to the 'REG 2' in the first column while other columns perform no operation ('NOP'). At the second cycle, the first column performs 'Execute1' from circular interconnection while PEs in the next column perform 'Load' from the first column. Then context words in the first registers are sequentially pipelined for two cycles (the third and forth cycle) and the first column performs 'Store' from the second register at the fifth cycle. Such a context pipelining is continually executed and finished at the eighth cycle. Therefore, if reusable context words are loaded into context registers in the circular order, the context words from spatial cache can be rotated for temporal mapping without temporal cache. It means that spatiality of the array structure and the added context registers can be utilized for low power in temporal mapping.

6.4.2   Limitation of Reusable Context Pipelining

If the loop given in Figure 6.6(b) is mapped onto the 4x4 PE array with added context registers like Figure 6.7(a), RCP cannot finish entire execution because the given architecture only supports a maximum number of 8 cycles (2 context registers and 4 columns) for an iteration of the loop whereas the loop has loop body taking 9 cycles. Therefore, in this case, temporal cache is necessary to support the entire execution as in Figure 6.9—RCP is performed for 4 cycles by register 1 and original context pipelining is performed for 5 cycles by register 2. Hence, RCP guarantees reduction of 4 cache-read operations after execution of the first iteration. This example shows that power efficiency of reusable context pipelining can be varied according to the complexity of evaluated loops and architecture specification.

Therefore, we can estimate how many cache-read operations occur after the first iteration under architecture constraints. This is given as follows:

$$N_{Tcache\_read} = 0 \qquad\qquad if \quad C_{iter} \leq m \times N_{ctxt} \qquad (6.2)$$

$$N_{Tcache\_read} = C_{iter} - m \times (N_{ctxt} - 1) \qquad if \quad C_{iter} > m \times N_{ctxt} \qquad (6.3)$$

where

- N_{Tcache_read}: cycle count of temporal cache-read operations after the first iteration

- C_{iter}: cycle count for an iteration of loop

- m: number of columns on reconfigurable array

- N_{ctxt}: number of context registers for a PE

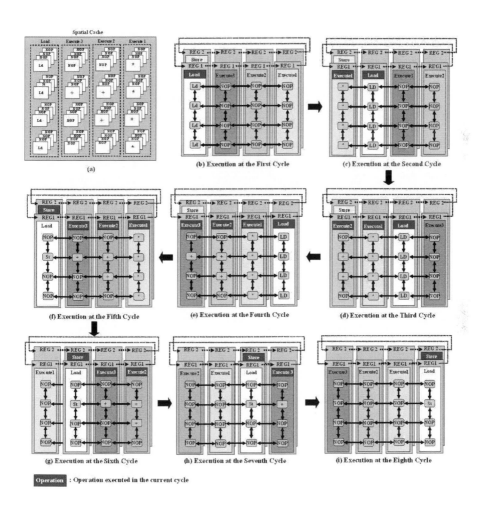

FIGURE 6.8: Reusable context pipelining for Eq. 6.1.

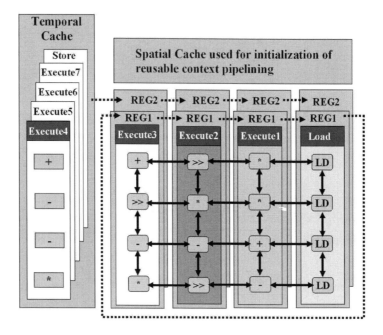

FIGURE 6.9: Reusable context pipelining with temporal cache.

	Cycle	i+1	i+2	i+3	i+4	i+5	i+6	i+7	i+8	i+9
Col#1	Cache/REG	REG1	REG1	REG1	REG1	Cache	Cache	Cache	Cache	Cache
	Operation	LD	EX1	EX2	EX3	EX4	EX5	EX6	EX7	ST

(a)

	Cycle	i+1	i+2	i+3	i+4	i+5	i+6	i+7	i+8
Col#1	Cache/REG	REG1	REG1	REG1	REG1	REG2	REG2	REG2	REG2
	Operation	LD	EX1	EX2	EX3	EX4	EX5	EX6	ST

(b)

FIGURE 6.10: Reusable context pipelining according to the execution time for one iteration (i > 1): (a) ith iteration in the case of loop body taking 9 cycles and (b) ith iteration in the case of loop body taking 8 cycles.

Based on above formula, the optimal case is when the $N_{T_{cache\_read}}$ is zero—context registers are sufficient to support entire loop body without temporal cache read-operations after the first iteration. Figure 6.10 shows two cases of temporal mapping with RCP after the first iteration. In the case of Figure 6.10(a), it shows the scheduling for previous example in Figure 6.9 and it corresponds to Eq. 6.3. However, Figure 6.10(b) shows other case that execution time for an iteration is 8 cycles and it corresponds to Eq. 6.2.

6.4.3 Hybrid Configuration Cache Structure

Based on modified interconnection structure as in Figure 6.7(b), we propose a power-conscious configuration cache structure that supports reusable context pipelining—we call it hybrid configuration cache including two cache parts—spatial cache for reusable context pipelining and temporal cache to make up for the limitation of RCP. Figure 6.11 shows the modified configuration cache structure to support the example given in Figure 6.9. It is composed of cache controller, spatial cache, temporal cache, multiplexer and demultiplexer. The cache controller supports the same functions as the previous controller and in addition it controls increased context registers as well as the selection between spatial cache and temporal cache. Therefore, the new cache controller is more complex than the base one but the cache controller supports reusable context pipelining with negligible area and power overheads. As compared to the distributed cache of base architecture, both spatial cache and temporal cache have much less number of layers since spatial mapping does not require many layers and RCP can save the layer of temporal cache by up to the number of columns using context registers—the number of spatial cache layers should be more than the number of context registers connected to a PE because spatial cache should be able to include context words of several applications. Therefore, the area cost overhead caused by added context registers offsets because temporal cache size can be reduced by same size of total added registers. As mentioned earlier, the approach does not incur any performance degradation and this hybrid structure saves cache area since we keep only one column with reduced number of temporal CEs and less layers of spatial CEs compared to distributed configuration cache that has much more layers of CEs.

6.5 Application Mapping Flow

We have implemented automatic compilation flow to map applications onto the base architecture for supporting temporal mapping [80]. The binary context words for reusable context pipelining are basically the same as the context words used for the temporal mapping but these context words should

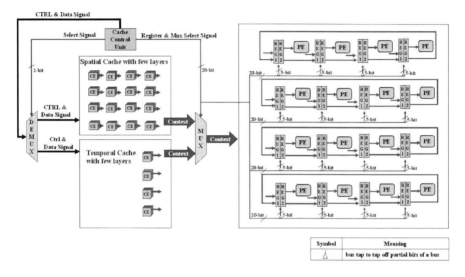

FIGURE 6.11: Hybrid configuration cache structure.

be rearranged for context pipelining with circular interconnection. Figure 6.12 shows entire mapping flow for the base architecture and proposed architecture. Binary context words are automatically generated from the compiler for temporal mapping. The timing and control information that is used to operate execution controller is manually optimized and the final encoded data is loaded onto registers of the execution controller.

6.5.1   Temporal Mapping Algorithm

The temporal mapping algorithm minimizes the execution time of kernel codes on the PE array. This execution time is directly proportional to the number of cache layers in configuration array. The time, $T_{critical}$ is considered as a parameter to be minimized during temporal mapping. We implement the temporal mapping in three sequential steps: covering, time assignment, and place assignment.

6.5.1.1   Covering

For compilation, the original kernel code is initially transformed into a DAG form, called the kernel DAG using common sub expression elimination technique [15]. One or more operation nodes in a kernel DAG are scheduled in a single configuration of a PE. For this, we generate a configuration DAG (CDAG) by clustering the nodes in kernel DAG. A CDAG is used to find the minimum number of configurations for kernel code execution. To perform this task, we formulate it into a DAG covering problem where one has to find the minimal cost set of patterns that cover all the nodes in input CDAG. To

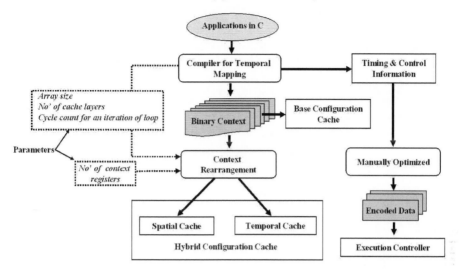

FIGURE 6.12: Application mapping flow for base architecture and proposed architecture.

efficiently solve our DAG covering problem, we implement our algorithm based on binate covering [36]. For example, Figure 6.13(a) shows CDAG generation from an input DAG.

6.5.1.2 Time Assignment

Each node in the CDAG is assigned to a cycle in which the node will be executed. In order to minimize $T_{critical}$, we must fully exploit the parallel resources provided by the m×n PE array using modulo scheduling [71]. For example, Figure 6.13(b) shows assignment schedule obtained after applying modulo scheduling to the CDAG. Note the cycle in which a node in the CDAG is scheduled as part of a configuration in this phase, and it represents a layer location inside a configuration cache.

6.5.1.3 Place Assignment

In this phase, we assign all nodes in the CDAG to actual PEs by storing each of them as a configuration entity in the cache of a PE. We split the PEs in a column into two groups, called slots. In this phase, the CDAG nodes are first assigned to either slot with resorting to the ILP solver, and then within each slot, nodes are finally mapped onto actual PEs. Figure 6.13(c) shows the final mapping results after a place assignment is deployed.

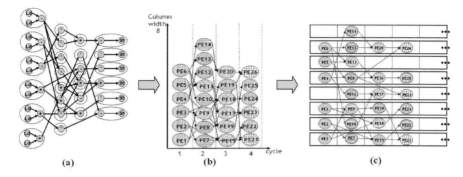

FIGURE 6.13: Temporal mapping steps: (a) covering; (b) time assignment; and (c) place assignment.

6.5.2   Context Rearrangement

In the case of base architecture, the binary context words generated from the compiler can be loaded into configuration cache without any modification. However, in the case of proposed architecture the generated context words are rearranged and properly assigned to spatial and temporal cache. The address of each context word in hybrid configuration cache can be represented by three-dimensional position as Figure 6.14(a). Figure 6.14(b) shows pseudo code for context rearrangement algorithm that is easily implemented based on Eq. 6.2 and 6.3. Before explaining the algorithm in detail, we introduce the notations used in the algorithm—$N_{T_{cache\_read}}$, C_{iter}, N_{ctxt} and m are defined in subsection C.2.

- n: number of rows in reconfigurable array

- k, l: number of temporal cache layers, the number of spatial cache layers

- T_{ctxt}: set of the context words having positions in temporal cache

- S_{ctxt}: set of the context words having positions in spatial cache

- (x, y, z): (layer index, row index, column index)

- $T_{ctxt}(x, y, z)$: context word corresponding the position (x, y, z) in temporal cache

- $S_{ctxt}(x, y, z)$: context word corresponding the position (x, y, z) in spatial cache

The code between L1 and L4 initialize temporary variables (p, r, u) and S_{ctxt}. If $N_{ctxt} \times m$ is sufficient to support entire loop body without temporal cache read-operations (L5), all of the context positions in temporal cache are remapped to the positions in spatial cache with rearrangement in the circular

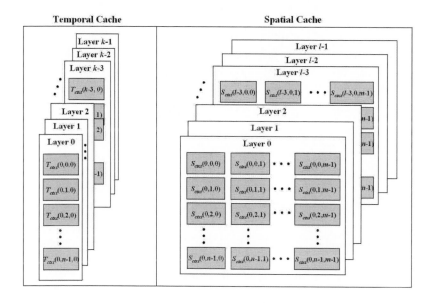

(a)

CONTEXT REARRANGEMENT (T_{ctxt}, m, n, k, C_{iter}, N_{ctxt})

```
L1    S_ctxt ← 0
L2    p ← 0
L3    r ← 0
L4    u ← 0
L5    if C_iter ≤ m×N_ctxt
L6        then for i ← 0 to k-1
L7            do for j ← 0 to n-1
L8                do S_ctxt(p, j, r) ← T_ctxt(i, j, 0)
L9                if r = 0
L10                   then r ← m - 1
L11                   else if r > 1
L12                       then r ← r - 1
L13                       else r ← 0, p ← p + 1
L14        else u ← m×(N_ctxt-1)
L15        for i ← 0 to k-1
L16            do if i ≤ u
L17                then for j ← 0 to n-1
L18                    do S_ctxt(p, j, r) ← T_ctxt(i, j, 0)
L19                    if r = 0
L20                       then r ← m - 1
L21                       else if r > 1
L22                           then r ← r - 1
L23                           else r ← 0, p ← p + 1
L24                else for j ← 0 to n-1
L25                    do T_ctxt(i-u-1, j, 0) ← T_ctxt(i, j, 0)
L26    return T_ctxt, S_ctxt
```

(b)

FIGURE 6.14: Context rearrangement: (a) positions of binary contexts in hybrid configuration cache and (b) rearrangement algorithm.

order (L6 ~ L13). Otherwise, the limited number of temporal cache layers which can be executed by reusable context pipelining is estimated (L14), all of the context positions within the limited temporal cache layers are remapped to the positions in spatial cache (L16 ~ L23) in the same manner as (L6 ~ L13). Then the layer indices of context positions remaining in temporal cache are updated to fill up the empty layers.

6.6   Experiments

6.6.1   Experimental Setup

For a fair comparison between the base model and the proposed one, we have implemented two cases of reconfigurable architectures as given in Table 6.1. Base architecture is as specified in Chapter 3. Proposed architecture is same as base architecture but also includes increased context registers and hybrid configuration cache to support reusable context pipelining. Two models have been designed at RT-level with VHDL and synthesized using Design Compiler [18] with 0.18 μm technology. We have used SRAM Macro Cell library for the frame buffer and configuration cache. ModelSim [17] and Prime-Power [18] have been used for gate-level simulation and power estimation respectively. To estimate the power consumption overhead in the proposed model, the context registers and multiplexers in each case (previous model and proposed architecture) have been separated from the PE array and those have been included in the configuration cache for each model. To obtain the power consumption data, we have used various kernels (Table 6.2) for simulation with same simulation conditions as the previous one mentioned in Chapter 3 (subsection 3.3)—operation frequency of 100 MHz and typical case of 1.8 V Vdd and 27°C. We have implemented the context rearrangement algorithm (Figure 6.14) in C++ and the application mapping flow as given in Figure 6.12 by adding the algorithm to the compiler for temporal mapping.

6.6.2   Results

6.6.2.1   Necessary Context Registers for Evaluated Kernels

We have applied several kernels of Livermore loops benchmark [6], DSP-stone [34] and representative loops in MPEG-4 AAC decoder, H.263 encoder and H.264 decoder to the base and proposed architectures. To determine necessary number of context registers to support reusable context pipelining for selected kernels, we have analyzed each case of selected kernels and Table 6.2 shows execution cycle count for an iteration and necessary number of context registers for each kernel. In the case of 2D-FDCT, it shows 11 execution cy-

TABLE 6.1: Architecture specification of base and proposed architecture

	Parameter	Base	Proposed
PE Array	No' of context registers for a PE	1	1
	No' of rows	8	8
	No' of columns	8	8
Frame Buffer	No' of sets and banks	2 and 3	2 and 3
	bit width	16-bit	16-bit
	Bank size	1KB	1KB
Configuration Cache	No' of layers for a CE	32	16
	No' of CEs	64	72
	Bit width of a CE	32-bit	32-bit

TABLE 6.2: Necessary context registers for evaluated kernels

Kernels	Execution cycle count for an iteration	Necessary number of context registers
[a]First_Diff	10	2
[a]Tri_Diagonal	4	1
[a]Hydro	7	2
[a]ICCG	5	1
[b]Dot_product	5	1
[b]24-Taps FIR	8	2
Complex Multiplication in MPEG-4 AAC decoder	10	2
ITRANS in H.264 decoder	9	2
2D-FDCT in H.263 encoder	11	2
SAD in H.263 encoder	5	1
Matrix(10x8)-Vector(8x1) Multiplication(MVM)	5	1

[a]Livermore loop benchmark suite. [b]DSP stone benchmark suite.

cles and the maximum number of context registers among selected kernels. It means that composing a PE having 2 context registers is necessary to support reusable context pipelining for all of the selected kernels. Therefore, each PE in the proposed architecture has 2 context registers for reusable context pipelining while base architecture has one context register as shown in Table 6.1.

6.6.2.2 Configuration Cache Size

Both temporal cache and spatial cache of the proposed architecture have 16 layers, which is half the size of the base architecture. Reducing cache size does not affect performance degradation of evaluated kernels—the size is sufficient to perform the selected kernels with reusable context pipelining. Table 6.3

TABLE 6.3:   Size of configuration cache and context registers

Size of memory elements	Architecture		Reduced(%)
	Base	**Proposed**	
Context registers	256-Byte	512-Byte	-
Spatial cache	8192-Byte	4096-Byte	43.75
Temporal cache		512-Byte	
Total amount	8448-Byte	5120-Byte	**39.39**

shows memory size evaluation between the base architecture and the proposed one. It shows that added context registers offset reduction of temporal cache layers. Compared to the base architecture, we have reduced the size of memory elements by up to 39.39%. This means that reconfigurable architecture with new configuration cache structure is more efficient than previous one in terms of memory size and power saving.

6.6.2.3   Performance Evaluation

The execution cycle counts of the evaluated kernels on proposed architecture do not vary from the base architecture because the functionality of proposed architecture is same as the base model. It also indicates the reusable context pipelining does not cause performance degradation in terms of the execution cycle count. In addition, the synthesis results show that the critical path delay of the proposed architecture is same as the base model i.e. 8.96 ns. It indicates the proposed approach does not cause performance degradation in terms of the critical path delay.

6.6.2.4   Power Evaluation

To demonstrate the effectiveness of our power-conscious approach, we have evaluated the power consumption of only base architecture with temporal mapping and proposed architecture with reusable context pipelining on hybrid configuration cache.

Table 6.4 shows comparison of power consumption between the two architectures. Selected kernels were executed with 100 iterations. Compared to the base architecture, we have saved up to 86.33% of the total power consumed in the configuration cache and 47.60 % of that in the entire architecture using reusable context pipelining. These results show that reusable context pipelining is a good solution for power saving in CGRA. ITRANS and SAD show less reduction in power compared to other kernels because they need additional spatial cache-read operations for data arrangement. In the case of 24-Taps FIR showing the maximum reduction ratio, the total power consumption of proposed architecture is much less than the result of PipeRench [74]. PipeRench has been fabricated in a 0.18 micron process and [74] shows power measurement with varying FIR filter tap sizes. The power consumption has been measured using a 33.3 MHz fabric clock and a 16.7 MHz IO clock. The power

TABLE 6.4: Power reduction ratio by reusable context pipelining

Kernels	Power(mW)				Reduced(%)	
	Cache		Entire		Cache	Entire
	Base	Proposed	Base	Proposed		
First_Diff	171.77	28.08	376.17	232.48	83.65	38.20
Tri_Diagonal	174.18	31.58	400.19	257.59	81.87	35.63
Dot_Product	117.84	29.87	328.54	240.57	74.65	26.78
Complex_Mult	180.63	32.82	452.00	304.19	81.83	32.70
Hydro	148.23	32.40	356.47	240.64	78.14	32.49
ICCG_Diff	205.80	32.64	434.45	261.29	84.14	39.86
24-Taps FIR	227.56	31.11	471.44	274.99	**86.33**	41.67
MVM	227.57	34.45	405.70	212.58	84.86	**47.60**
ITRANS	204.85	69.96	417.95	283.06	65.85	32.27
2D-FDCT	190.03	37.59	417.33	264.89	80.22	36.53
SAD	185.30	75.08	415.27	305.05	59.48	26.54

measurement shows that the power consumption of 24-Taps FIR ranges from 600 mW to 700 mW.

6.7 Summary

Most reconfigurable architectures have a configuration cache for dynamic reconfiguration, which consumes high amount of power. In this chapter, we introduce reusable context pipelining (RCP) for low-power reconfiguration and hybrid configuration cache structure supporting this technique. RCP can be used to achieve power-savings in a reconfigurable architecture while maintaining performance same as general CGRA. In addition, the hybrid configuration cache structure is more efficient than previous one in terms of memory size. In the experiments, we show that the proposed approach saves power even with reduced configuration cache size. Power reduction ratios in the configuration cache and the entire architecture are up to 86.33% and 47.60% respectively compared to the base architecture.

Chapter 7

Dynamic Context Compression for Low-Power CGRA

7.1 Introduction

Most of the coarse-grained reconfigurable array architectures (CGRAs) are composed of reconfigurable ALU arrays and configuration cache (or context memory) to achieve high performance and flexibility. Specially, configuration cache is the main component in CGRA that provides distinct feature for dynamic reconfiguration in every cycle. However, frequent memory-read operations for dynamic reconfiguration cause much power consumption. Thus, reducing power in configuration cache has become critical for CGRA to be more competitive and reliable for its use in embedded systems. In this chapter, we address the power reduction issues in CGRA and provide a framework to achieve this. A design flow and a configuration cache structure are presented to reduce power consumption in configuration cache [41]. The power saving is achieved by dynamic context compression in the configuration cache—only required bits of the context words are set to enable and the redundant bits are set to disable. Therefore, the efficient design flow for CGRA has been proposed to generate architecture specifications that are required for supporting dynamically compressible context architecture without performance degradation. Experimental results show that the proposed approach saves up to 39.72% power in configuration cache with negligible area overhead (2.16%).

7.2 Preliminary

7.2.1 Context Architecture

The configuration cache provides context words to the context register of each PE on a cycle by cycle basis. From the context register, these context words configure the PEs. Figure 2.4 shows an example of PE structure and context architecture for MorphoSys [75]. 32-bit context word specifies the function for the ALU-multiplier, the inputs to be selected from MUX_A and

MUX_B, the amount and direction of shift of the ALU output, and the register for storing the result as Figure 2.4(a) in Chapter 2. Context architecture means organization of context word with several fields to control resources in a PE as Figure 2.4(b) in Chapter 2.

The context architectures of other CGRAs such as [9, 11, 19, 21, 23, 26, 29, 37, 40, 46, 62, 76] are similar to the case of MorphoSys although there is a wide variance in context-width and kind of fields used by different functionality.

7.3   Motivation

7.3.1   Power Consumption by Configuration Cache

By loading the context words from the configuration cache into the array, we can dynamically change the configuration of the entire array within just one cycle. However, such dynamic reconfiguration of CGRA causes many SRAM-read operations in configuration cache. In [74], the authors have fabricated a CGRA (PipeRench) in a 0.18 μm process. Their experimental results show that the power consumption is significantly high due to the dynamic reconfiguration requiring frequent configuration memory access. In Figure 5.1, power break-down for the CGRA running 2D-FDCT is proposed with gate-level implementation at 0.18 μm technology based on MorphoSys architecture. It is shown that the configuration cache spends about 43% of the overall power, which is the second largest after the PE arrays consuming 48% of overall power budget. This is because the configuration cache performs SRAM-read operations to load the context words in every cycle at run time. In addition, [21,48] also shows power break-down for another CGRA (ADRES) running IDCT based on 90nm technology. In this case, the configuration memory spends about 37.22% of the overall power. Therefore, it is explicit that power consumption by configuration cache (memory) is serious overhead compared to other types of IP cores such as ASIC or ASIP.

7.3.2   Valid Bit-Width of Context Words

When a kernel is mapped onto CGRA and application gets executed, the usable context fields are limited to types of operations involved due to the kernel executed at run time. Furthermore, operation types of an executed kernel on PE array are changed in every cycle. It means the valid bit-width of executed context word is frequently less than the full bit-width of a context word even though full bit-width can be less often used. For statistical evaluation of valid bit-width of contexts, we selected 32-bit context architecture of the base architecture (Figure 4.3) and mapped several kernels onto its PE array in order to maximize the utilization of the context fields. Figure 7.1 shows the

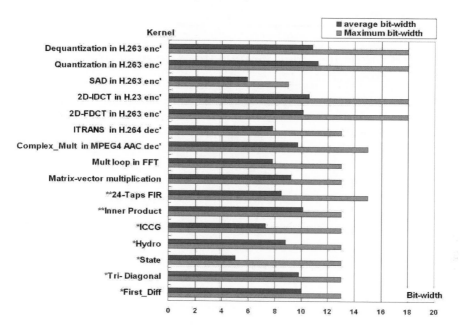

FIGURE 7.1: Valid bit-width of context words: *Livermore loops benchmark [6], **DSPstone [34].

results for various benchmark kernels and critical loops in real applications. In Figure 7.1, average bit-width is the average value of valid bit-widths of all the executed context words at run-time and the maximum bit-width is the maximal valid bit-width among all the context words considered at run-time. The statistical result shows that average bit-widths vary from 7 to 11 bits and the maximum bit-width is less than or equal to 18 bits whereas the full bit-width is 32-bit.

7.3.3 Dynamic Context Compression for Low-Power CGRA

If the configuration cache can provide only required bits (valid bits) of the context words to PE array at run time, it is possible to reduce power consumption in configuration cache. The redundant bits of the context words can be set to disable and make those in-valid at run time. That way, one can achieve low-power implementation of CGRA without performance degradation while context architecture dynamically supports both the cases at run time: one case is uncompressed context word with full bit-width and another case is compressed context word with setting unused part of configuration cache disabled. In order to support such a dynamic context compression, we propose a new context architecture and configuration cache structure in this chapter.

7.4   Design Flow of Dynamically Compressible Context Architecture

In order to design and evaluate dynamically compressible context architecture, we propose a new context architecture design flow. Entire design flow is shown in Figure 7.2. This design starts from context architecture initialization, which is similar to the architecture specification stage of general CGRA design flow given in [7, 48, 56, 60]. Based on such architecture specifications, PE operations are determined and initial context architecture is defined. From the context initialization, fields are grouped by essentiality of PE operation and dependency with ALU operation to provide some criterions for context compression. A field sequence graph (FSG) is generated to show possible field combinations for PE operation. Then field control signals are generated to make some field enable or disable when contexts are compressed. Based on former stages, the position of each field is defined and final context architecture is generated. Finally, one can determine whether the initially uncompressed contexts can be compressed or not by context evaluator. From subsection 7.4.1 to subsection 7.4.5, we describe more detailed process for each stage in entire design flow.

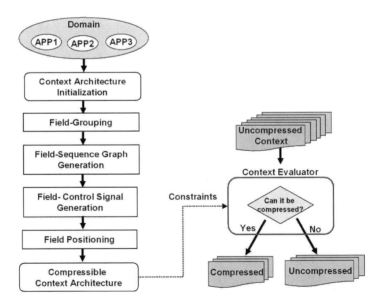

FIGURE 7.2: Entire design flow.

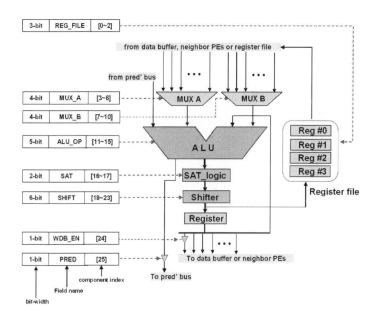

(a)

Field name	Bit-width	Component index	Control
SHIFT	6-bit	[18,19,20,21,22,23]	
ALU_OP	5-bit	[11,12,13,14,15]	
MUX_A	4-bit	[3,4,5,6]	
MUX_B	4-bit	[7,8,9,10]	Processing
REG_FILE	3-bit	[0,1,2]	Element
SAT	2-bit	[16,17]	
WDB_EN	1-bit	[24]	
PRED	1-bit	[25]	
CTXT_CTRL	6-bit	-	Context register

(b)

FIGURE 7.3: Context architecture initialization: (a) PE structure; (b) context architecture initialization.

7.4.1   Context Architecture Initialization

Context architecture in CGRA design depends on architecture specification. In the process of architecture specification, CGRA structure is evolved with PE array size, PE functionalities and their interconnect scheme. The proposed approach starts from the conventional context architecture selection and makes it dynamically compressible context architecture through the proposed design flow. We have defined generic 32-bit context architecture as an example to illustrate the design flow to support the kernels in Figure 7.1. It is similar to the representative CGRAs such as MorphoSys [75], REMARC [61], ADRES [7,21,78,82], PACT_XPP [11,28,29]. The PE structure and bit-width of each field are shown in Figure 7.3. It supports various arithmetic and logical operations with two operands (MUX_A and MUX_B), predicated execution (PRED), Arithmetic saturation (SAT_logic), shift operation (SHIFT) and saving temporal data with register file (REG_FILE). In Figure 7.3(a), all of the fields are classified by 'Control' of 2 cases—'Processing element' and 'context register'. It means that each case is configured by the fields included in that case. Furthermore, Figure 7.3(b) shows the bit-width of each field and the component index to identify each component configured by each field. Even though each field can be positioned on context word under conventional design flow, this initialization stage does not define any field position. It means field position for uncompressed case should be assigned by considering context compression.

7.4.2   Field Grouping

All of the context fields are grouped into three sets—necessary set, optional set and unnecessary set. Necessary set includes indispensable fields for all of the PE operations and optional set includes optional fields for PE operations. Unnecessary set is composed of fields unrelated to PE operations. It means necessary fields should be included in context words even if context words are compressed whereas optional and unnecessary fields can be excluded out of context words. In addition, we classify optional set into two subsets. One is a subset composed of fields dependent on the field of 'ALU_OP' and another is a subset composed of fields independent of 'ALU_OP'. This classification is necessary for generating field control signals in subsection 7.4.4. Figure 7.4 shows field grouping based on the context initialization presented in subsection 7.4.1.

7.4.3   Field Sequence Graph Generation

Field sequence graph (FSG) is generated from context architecture initialization and field grouping. FSG is a directed graph composed of necessary and optional fields and it shows possible field combinations for PE operations based on PE structure. Each vertex of FSG corresponds to a necessary or

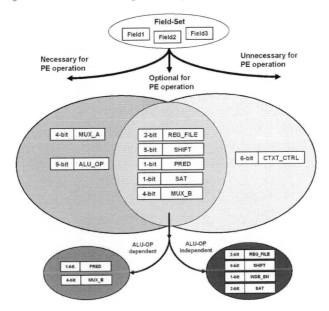

FIGURE 7.4: Field grouping.

optional field in field grouping and each edge of FSG shows a possible field combination between two fields. The possible field combinations can be found by vertex tracing in the edge directions and the combinations should include all of the necessary fields. Furthermore, optional fields can be skipped out of vertex tracing to search possible field combinations. Figure 7.5 shows an example of FSG from Figure 7.3 and Figure 7.4. While searching possible field combinations, sometimes it is possible (for example, MUX_A, ALU_OP, SAT is possible whereas MUX_A, ALU_OP, SAT, PRED is not possible). FSG is a useful data structure for field positioning as described in subsection 7.4.5.

7.4.4 Generation of Field Control Signal

When contexts are compressed, optional fields are relocated on compressed space and the positions of these fields may be overlapped with each other. Therefore, each optional field should be disabled when it is not being compressed in the context word. It means that compressed context should have control information for all of the optional fields in order to make unused fields disable. In this subsection, control signals generation for optional fields has been described.

7.4.4.1 Control Signals for ALU-Dependent Fields

If the truth table of 'ALU_OP' is classified by the operation type, enable/disable signals for ALU-dependent fields can be generated from

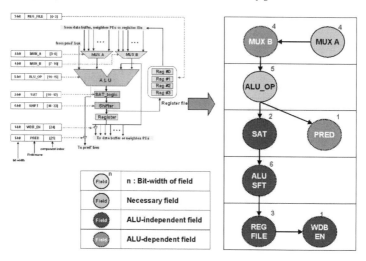

FIGURE 7.5: Field sequence graph.

'ALU_OP' with some combinational logic. Figure 7.6(a) shows the truth table manipulated by classifying operations for the example given in subsection 7.4.1. MSB (A4) of 'ALU_OP' is used for classifying operations according to the number of operands. For example, MSB =1 is used for the operations with two operands and MSB =0 is used for the operations with one operand. In addition, A3 ∼ A0 are used for classifying logical operations. Based on the truth table, we can generate control signals for two fields with some combinational logic as Figure 7.6(b). We define such a combinational logic as 'CTRL BLOCK'.

7.4.4.2   Control Signals for ALU-Independent Fields

In order to control ALU-independent fields when context words are compressed, the enable/disable flag bit on each of the ALU-independent fields should be merged with a necessary field. Figure 7.7(a) shows the process that 1-bit flags of ALU-independent fields are merged with 'ALU_OP'. After flag merging, the FSG should be updated because the bit-widths of some of the fields are changed and 1-bit field such as 'WDB_EN' is no longer valid in FSG. Figure 7.7(b) shows an updated FSG with modified bit-widths of some of the fields.

7.4.5   Field Positioning

The final stage of proposed design flow is positioning each field on the context word. Field positioning should be considered for two cases (uncompressed and compressed modes) to support dynamic compression.

Logical Operation	ALU_OP [5-bit]				
	A_4	A_3	A_2	A_1	A_0
A && B	1	1	1	0	0
A || B	1	1	1	0	1
A < B	1	1	1	1	0
A ≤ B	1	1	1	1	1
A!	0	1	1	1	1

(a)

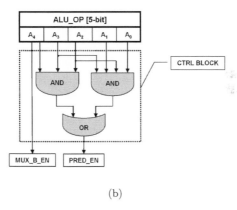

(b)

FIGURE 7.6: Control signals for 'MUX_B' and 'PRED': (a) logical operations; (b) control signals.

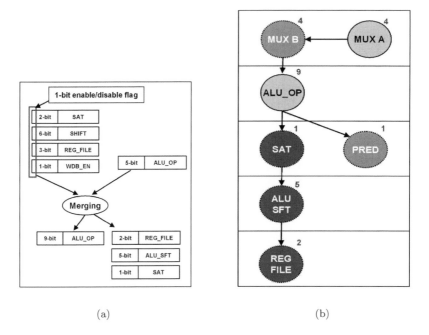

(a)   (b)

FIGURE 7.7: Updated FSG from flag merging: (a) flag merging; (b) updated FSG.

7.4.5.1   Field Positioning on Uncompressed Context Word

All the fields should have default positions for the case when contexts cannot be compressed. First of all, the necessary fields are positioned to the part near to MSB and the unnecessary fields are positioned near the LSB as shown in Figure 7.8. Then the optional fields are positioned on the available space between the already occupied context word. For optional field positioning, the bit-width of compressed context word should be determined. Compressed bit width can be different according to the definition of the capacity of compressed context word. The large capacity of compressed context word can show high compression ratio but the amount of power reduction is limited by long bit-width. However, the little capacity of compressed context word may cause low compression ratio but the power reduction ratio can be high in short bit-width. To prevent the extreme cases (much short or much long bit-width of compressed context word), we determine compressed bit-width based on the following criterions.

- *i*) Compressed context words should be able to support all of the ALU-dependent fields.

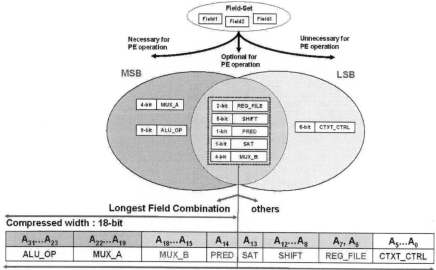

FIGURE 7.8: Default field positioning.

- *ii*) Compressed context words should be able to include at least an ALU-independent field.

To satisfy criterions, we determine the longest field combination showing the maximum bit-width among i) and ii). The maximum width for satisfying i) and ii) is found to be 18-bit that consists of 'ALU_OP', 'MUX_A', 'MUX_B' and 'PRED'. Therefore, 18-bit is the compressed bit-width. Optional fields that are included in the longest field combination are preferentially positioned on the compressed zone near the MSB and other fields are positioned on uncompressed zone near the LSB as Figure 7.8. After this, the positions of the necessary fields on FSG are firmly determined and the positions of the field control signals are also determined because they are included in 'ALU_OP' as necessary field.

7.4.5.2 Field Positioning on Compressed Context Word

This stage is for positioning fields on compressed context word to guarantee that all the possible field combinations are not exceeding the compressed bit-width. Therefore, first of all, all the possible field combinations should be found. This process can be achieved by searching them from FSG and then generating field concurrency graph (FCG) such as Figure 7.9(a). The FCG shows the concurrency between the optional fields. Therefore the FCG is used for preventing position that is overlapping between the concurrent optional fields. An edge between two fields means that the two fields are included in one of the possible field combinations. Even though this example does not show

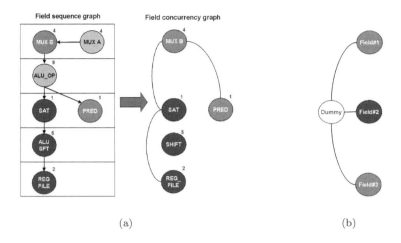

(a)   (b)

FIGURE 7.9: Field concurrency graph: (a) FCG from FSG; (b) CG with dummy vertex.

concurrency among more than 2 optional fields, such a case can be represented by adding a dummy field connected with the fields as Figure 7.9(b). Based on a given FCG, the next step is to position the optional fields on compressed context word. The positioning means that some optional fields have additional positions as well as default positions on uncompressed context words. To select a position among default and additional positions, multiplexers can be used that are composed of multiple position inputs and one feasible position output. Therefore, in this step, the field positioning is a mapping among inputs, outputs and control signals for multiplexers connected with the optional fields. Thus, we propose a port-mapping algorithm for the multiplexers. Before we explain the procedure in detail, we introduce notations we use in the explanation as Table 7.1.

The input to the port-mapping algorithm is FCG and the output is multiplexer port-mapping graph (PMG) showing the relationship among field control signals and input data signals (field position). The algorithm is composed of two parts—the first part is for the optional fields having default position on compressed context word and the second part is for the optional fields not having default position on compressed context word. The procedure of the first part is described in Algorithm 1. The algorithm starts with initialization step (L1 and L2). In this part, input data signals of multiplexers are only two cases—default field position and 'zero' selected when the field is not used. This is because the fields already have default positions on compressed context space. Therefore the default field position, 'zero' and the field control signal of each field are mapped to the input of the multiplexer (L4 ∼ L6). Next process

TABLE 7.1: Notations for port-mapping algorithm

Notation	Meaning
G_{FCG}	field concurrency graph, $G_{FCG} = (V, E)$: V is a set composed of the optional field set and E is a set composed of edges showing the concurrency between two fields.
G_{MUX}	multiplexer port mapping graph $G_{MUX} = (V_{MUX}, E_{MUX})$: V_{MUX} is a set composed of input signals and control signals for multiplexers and E_{MUX} is a set composed of weighted edges connecting input data with control signal.
$defV$	subset of V, $defV$ is composed of the fields having their default positions on compressed context word
$ndefV$	subset of V, $ndefV$ is composed of the fields not having their default positions on compressed context word
$ctxt[A_i, A_j]$	bit interval from index A_i to index A_j on the uncompressed context word, it is used for showing bit position of a field.
$width[v]$	bit-width of field v
cmp_lsb	LSB of compressed context word
$def\_pos[field]$	default position of field such as interval type of $ctxt[A_i, A_j]$
$ctrl\_pos[field]$	one bit position of control signal for field such as $ctxt[A_i]$, $ctrl_blk[A_i]$
$field[i, j]$	component index corresponding to the interval that is from the i_{th} bit position to the j_{th} bit position on field
cmp_ctrl	one-bit signal from cache control unit. '1' means executed context word compressed and '0' means executed context word not compressed.
$pdone[field]$	'1' means positioning firmly done and '0' means positioning not finished.
$mux[field]$	mux(multiplexer) connected with field.
$data\_in[mux]$	set composed of mux input data signals
$ctrl\_in[mux]$	set composed of field control signals for mux
$data\_out[mux]$	set composed of mux ouput data signals
$Adj[field]$	adjacency list of $field$ on graph G_{FCG}, if an adjacent field is dummy, it return adjacency list of the dummy field

Algorithm 1 Mux_Port Mapping (G_{FCG}) - fields having default position	
L1	$V_{MUX} \leftarrow \varnothing, E_{MUX} \leftarrow \varnothing, G_{MUX} \leftarrow (V_{MUX}, E_{MUX})$
L2	Add cmp_ctrl on V_{MUX}
L3	**for** each $v \in defl$ **do**
L4	$data\_in[mux[v]] \leftarrow data\_in[mux[v]] \cup \{def\_pos[v]\} \cup \{[null]\}$
L5	$ctrl\_in[mux[v]] \leftarrow ctrl\_in[mux[v]] \cup \{ctrl\_pos[v]\}$
L6	Add $data\_in[mux[v]]$ and $ctrl\_in[mux[v]]$ to V_{MUX}
L7	Add an edge between $def\_pos[v]$ and $ctrl\_pos[v]$ with weight '1' to E_{MUX}
L8	Add an edge between $[null]$ and $ctrl\_pos[v]$ with weight '0' to E_{MUX}
L9	$data\_out[mux[v]] \leftarrow v[width[v]-1, 0]$
L10	$pdone[v] = 1$
L11	**end do**

is to define the relationship between field control signal and a field position by adding a weighted edge between them (L7 and L8). Weight '1' (or '0') means the input signal is selected when the control signal is '1' (or '0'). Finally, the outputs of multiplexers are connected with the component index defined in subsection 7.4.1(L9) and positioning of the field is firmly done (L10).

The procedure of the second part is described in Algorithm 2. The algorithm starts with mapping default field position and signal 'cmp_ctrl' to the input of the multiplexer for each field (L2 and L3). Signal 'cmp_ctrl' is one-bit signal from cache control unit and it gives information whether the context word is compressed ('1') or not ('0'). Then the algorithm defines the relationship between signal 'cmp_ctrl' and a default position by adding a edge showing weight '0' between them (L5). Next process is split into two cases— one is for the fields having no adjacent fields on FCG and another is for the fields having adjacent fields on FCG. The first case means the fields can be positioned to any part of compressed zone except the positions of necessary fields whereas the second case means the fields should be positioned to the part not overlapped with the positions of their adjacent fields. In the first case (L6), the field is positioned to the part near to LSB of compressed context word (L7). Then new field position and field control signal are mapped to the input of the multiplexer (L8 and L9). Next process is to define the relationship between field control signal (or 'cmp_ctrl') and new field position by adding a edge showing weight '1' between them (L11 and L12).

In the second case (L13), 'Check_Adjancency' function is used and it is described as Algorithm 3. The algorithm starts with gathering the adjacent fields firmly positioned. Then new position on compressed zone is assigned by 'Find_Interval' function (L6). Figure 7.10 shows examples for this function with two cases—(c) when new position of input field is not overlapped with the adjacent field positions and (d) when new position of input field is overlapped with the adjacent field positions. 'Find_Interval' only returns a new position (ctxt[Ai, Aj]) in Figure 7.10(c) because of no confliction with the adjacent

Algorithm 2 Mux_Port Mapping (G_{FCG}) - fields not having default position	
L1	**for** each $v \in ndefV$ **do**
L2	$data\_in[mux[v]] \leftarrow data\_in[mux[v]] \cup \{def\_pos[v]\}$
L3	$ctrl\_in[mux[v]] \leftarrow ctrl\_in[mux[v]] \cup \{cmp\_ctrl\}$
L4	Add $def\_pos[v]$ to V_{MUX}
L5	Add an edge between $def\_pos[v]$ and cmp_ctrl with weight '0' on E_{MUX}
L6	**if** $Adj[v] = \emptyset$ on G_{FCG} **then**
L7	$tmp\_interval \leftarrow ctxt[(width[v]+cmp\_lsb), cmp\_lsb]$
L8	$data\_in[mux[v]] \leftarrow data\_in[mux[v]] \cup \{tmp\_interval\}$
L9	$ctrl\_in[mux[v]] \leftarrow ctrl\_in[mux[v]] \cup \{ctrl\_pos[v]\}$
L10	Add $data\_in[mux[v]]$ and $ctrl\_in[mux[v]]$ to V_{MUX}
L11	Add an edge between $tmp\_interval$ and $ctrl\_pos[v]$ with weight '1' to E_{MUX}
L12	Add an edge between $tmp\_interval$ and cmp_ctrl with weight '1' to E_{MUX}
L13	**else** Check_Adjacency(v)
L14	**end if**
L15	$data\_out[mux[v]] \leftarrow v[width[v]-1, 0]$
L16	$pdone[u] \leftarrow 1$
L17	**end do**

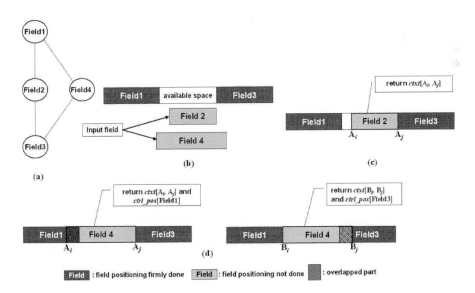

FIGURE 7.10: Examples of 'Find_Interval': (a) FCG; (b) available space on compressed zone; (c) when target field ('Field2') not overlapped with adjacent fields; (d) when target field ('Field4') overlapped with adjacent fields ('Field1' or 'Field3').

Algorithm 3 Check_Adjacency (*field*)

L1	**for each** $u \in Adj[v]$ on G_{FCG} **do**
L2	**if** $pdone[u] = 1$ **then**
L3	$tmpI' \leftarrow tmpI' \cup \{u\}$
L4	**end if**
L5	**end do**
L6	*position_set* and *ctrl_set* $\leftarrow$ Find_Interval (v, $tmpV$)
L7	**for each** $ctxt[A_i, A_j] \in position\_set$ **do**
L8	**if** $\{ctxt[A_i, A_j]\} \cap data\_in[mux[v]] = \emptyset$ **then**
L9	$data\_in[mux[v]] \leftarrow data\_in[mux[v]] \cup \{ctxt[A_i, A_j]\}$
L10	$ctrl\_in[mux[v]] \leftarrow ctrl\_in[mux[v]] \cup \{ctrl\_pos[v]\} \cup ctrl\_set$
L10	Add $data\_in[mux[v]]$ and $ctrl\_in[mux[v]]$ to V_{MUX}
L11	Add an edge between $ctxt[A_i, A_j]$ and $ctrl\_pos[v]$ with weight '1' to E_{MUX}
L12	Add an edge between $ctxt[A_i, A_j]$ and cmp_ctrl with weight '1' to E_{MUX}
L13	**for each** $ctrl \in ctrl\_set$ **do**
L14	**if** $ctrl$ overlapped with $ctxt[A_i, A_j]$ **then**
L15	Add an edge between $ctxt[A_i, A_j]$ and $ctrl$ with weight '0' to E_{MUX}
L16	**end if**
L17	**end do**
L18	**end if**
L18	**end do**

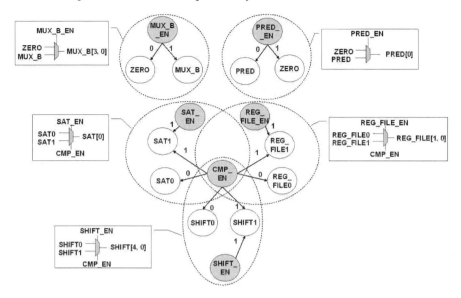

FIGURE 7.11: Multiplexer port-mapping graph.

fields. However, it returns two positions (ctxt[Ai, Aj] and ctxt[Bi, Bj]) and field control signals from overlapped fields in Figure 7.10(d). This is because the adjacent field control signals are necessary to select a proper field position when multiple field positions exist on compressed zone. Such returned new position set and control signal set are mapped to the input of multiplexer for the input field (L9 and L10) and the relationship among field control signals and a new position is made by adding weighted edges among them (L11 ∼ L17). Finally, the outputs of multiplexers are connected with the component index (L15) and positioning of the field is firmly done (L16) in Algorithm 2.

PMG example from the port-mapping algorithm is shown in Figure 7.11. Each vertex of PMG corresponds to an input or control signal of multiplexer and each edge shows the relationship between control signal and a position that is selected by the weight of the edge from control signals such as 'SAT_EN', 'MUX_B_EN', etc. Then the outputs of multiplexers are connected with the component index defined in Figure 7.3(b). Therefore we can implement the multiplexers for the optional fields by the PMG.

7.4.6 Compressible Context Architecture

After the field positioning, we have generated a specification of dynamically compressible context architecture like the one in Figure 7.12. Figure 7.12(a) shows the final field layout of compressible context architecture. 'REG_FILE', 'SHIFT' and 'SAT' have double positions for compressed and uncompressed cases. Figure 7.12(b) shows a modified structure between a PE and a cache

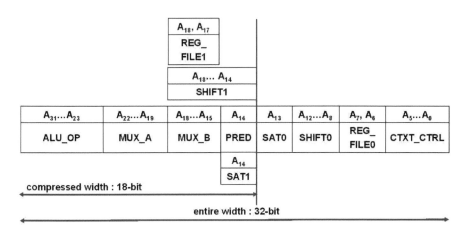

A₃₁...A₂₃	A₂₂...A₁₉	A₁₈...A₁₅	A₁₄	A₁₃	A₁₂...A₈	A₇, A₆	A₅...A₀
ALU_OP	MUX_A	MUX_B	PRED	SAT0	SHIFT0	REG_FILE0	CTXT_CTRL

(a)

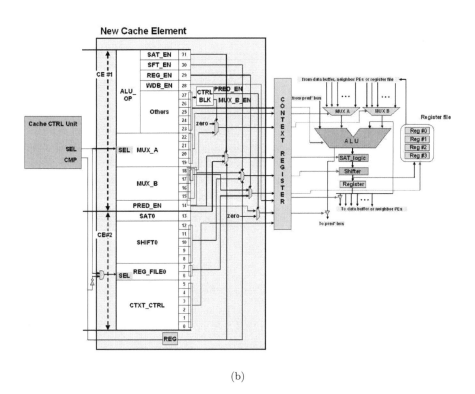

(b)

FIGURE 7.12: Compressible context architecture: (a) field layout of compressible context architecture; (b) modified structure between a PE and a CE.

element (CE). New cache element is composed of CE1 and CE2 and cache control unit provides compression information from port 'CMP' whether executed contexts are compressed or not. CE1 is always selected but CE2 is not selected under compression ('CMP'=1) to remove power consumption in CE2.

7.4.7   Context Evaluation

The context evaluator in Figure 7.2 determines whether initially uncompressed contexts can be compressed or not. This evaluation process can be implemented by checking the fact that a given context word is compared with one of the possible field combinations not exceeding compressed bit-width. Using FCG, we can easily check this and generate compressed context words with using position information from PMG.

7.5   Experiments

7.5.1   Experimental Setup

We have implemented entire design flow in Figure 7.2 with C++. We have initialized context architecture as the example described in Section 7.4. The implemented design flow generated the specification of dynamically compressible context architecture. For quantitative evaluation, we have designed two CGRAs based on the 8x8 reconfigurable array at RT-level with VHDL—one is conventional base CGRA and the other is the proposed CGRA supporting compressible features in context architecture. The architectures have been synthesized using Design Compiler [18] with 0.18 μm technology. We have used SRAM Macro Cell library for the frame buffer and configuration cache. ModelSim [17] and PrimePower [18] tools have been used for gate-level simulation and power estimation. To obtain the power consumption data, we have used the kernels (Figure 7.1) for simulation with operation frequency of 100 MHz and typical case of 1.8 V Vdd and 27°C. These kernels have been executed with 100 iterations while varying test vectors.

7.5.2   Results

7.5.2.1   Area Cost Evaluation

Table 7.2 shows the synthesis results from Design Compiler [18] of proposed architecture and base architecture. It shows that area cost of new configuration cache including cache control unit, added interconnects and multiplexers has increased by 10.35% but the overall area-overhead is only 1.62%. Thus, the

TABLE 7.2:   Area overhead by dynamic context compression

Component	Area Cost (gate equivalent)		Overhead (%)
	Base	Proposed	
Configuration Cache	150012	165538	10.35
Entire RAA	942742	958268	1.62
Overhead (%): {(Proposed/Base) - 1}×100			

TABLE 7.3:   Power reduction ratio by dynamic context compression

Kernels	Compression Ratio (%)	Configuration Cache Power(mW)		Reduced (%)
		Base	Proposed	
First_Diff	100	171.77	104.97	38.89
Tri-Diagonal	100	174.18	105.00	39.72
State	100	161.23	99.38	38.36
Hydro	100	148.23	91.50	38.27
ICCG	100	205.80	125.68	38.93
Inner Product	100	117.84	72.60	38.39
24-Taps FIR	100	227.56	139.56	38.67
MVM	100	227.57	140.43	38.29
Mult in FFT	100	175.48	107.08	38.98
Complex Mult	100	180.63	110.18	39.00
ITRANS	100	204.85	125.27	38.85
2D-FDCT	95.53	190.03	119.87	36.92
2D-IDCT	95.49	188.47	118.98	36.87
SAD	100	185.30	113.07	38.98
Quant	95.12	185.23	117.51	36.56
Dequant	95.23	187.78	118.77	36.75
Compression Ratio (%): (number of compressed context words/				

Compression Ratio (%): (number of compressed context words/
                                number of entire context words)×100,
Reduced (%): {1-(Proposed/Base)}×100,
Execution Cycle Count : cycle count for an iteration.

new configuration cache structure can support dynamic context compression with negligible overheads.

7.5.2.2   Performance Evaluation

In addition, the synthesis results show that the critical path delay of the proposed architecture is same as the base model i.e. 8.96 ns. It indicates the dynamic context compression does not cause performance degradation in terms of the critical path delay. In addition, we have applied several kernels in Figure 7.1 to the new and base architectures. The execution cycle count of each kernel on proposed architecture does not vary from the base architecture because the functionality of proposed architecture is same as the base model.

It also indicates the dynamic context compression does not cause performance degradation in terms of the execution cycle count.

7.5.2.3   Context Compression Ratio and Power Evaluation

Table 7.3 shows context compression ratio for the evaluated kernels. Compression ratio means how many context words can be compressed among entire context words. The execution cycle count of each kernel on proposed architecture does not vary from the base architecture because the functionality of proposed architecture is same as the base model. It also indicates the dynamic context compression does not cause performance degradation in terms of the execution cycle count. All of the kernels show high compression ratio to be more than 95%. Furthermore, the comparison of power consumption is shown in Table 7.3. Compared to the base architecture, it has shown to save up to 39.72% of the power. Four kernels (2D-FDCT, 2D-IDCT, Quant and Dequant) show less reduction in power compared to other kernels. This is because all of the context words for 4 kernels are not fully compressed—the compression ratios are in the range of 95.12 ∼ 95.53.

7.6   Summary

Power consumption is very crucial for the coarse-grained reconfigurable architecture for embedded systems and all reconfigurable architectures have a configuration cache for dynamic reconfiguration, which consumes significant amount of power. In this chapter, we introduce dynamically compressible context architecture with its design flow and configuration cache structure to support it. The dynamically compressible context architecture can save power in configuration cache without performance degradation. Experimental results show that the approach saves much power (by up to 39.72%) compared to conventional base model with negligible area overhead.

Chapter 8

Dynamic Context Management for Low-Power CGRA

8.1 Introduction

The configuration cache is the main component in CGRA that provides distinct feature for dynamic configuration. Even though configuration cache plays an important role for high performance and flexibility, it suffers from large power consumption. Therefore, reducing power consumption in the configuration cache has been a serious concern for reliability of CGRA. In this chapter, we present a control mechanism of configuration cache called dynamic context management to reduce the power consumption in configuration cache without performance degradation [44]. In addition, an efficient configuration cache structure is introduced to support such a dynamic context management. Experimental results show that the proposed approach saves 38.24%/38.15% of the power in write/read-operation of configuration cache with negligible area overhead compared to the base design.

8.2 Motivation

8.2.1 Power Consumption by Configuration Cache

As mentioned in 7.3.1, the power consumption in CGRA is significantly high due to the dynamic reconfiguration requiring frequent configuration memory access. It means that power consumption by configuration cache (memory) is serious overhead compared to other types of IP cores such as ASIC or ASIP.

8.2.2 Redundancy of Context Words

Context words are saved in configuration cache and they show redundancies at runtime. We describe two cases for redundancy of context words in the following subsections.

8.2.2.1   NOP Context Words

Most coarse-grained reconfigurable arrays arrange their processing elements (PEs) as a square or rectangular 2-D array with horizontal and vertical connections, which support rich communication resources for efficient parallelism. However, such PE arrays have many redundant or unutilized PEs during the executions of applications onto the array. Most subtasks in DSP applications show lots of redundant PEs that are not used. The redundant PEs should be configured by NOP (no operation) context words to avoid malfunction and unnecessary waste of power by the PEs. It means that configuration cache performs some redundant read-operations for NOP.

8.2.2.2   Consecutively Same Part in Context Words

When a kernel is mapped onto CGRA and application gets executed, the consecutively changed context fields are limited to types of operations involved due to the kernel executed at run time. Figure 8.1 shows 3 cases for consecutively same part in context words at run time. In the case of Figure 8.1(a), PEs perform continuous 'Load' operations with fixed 'ALU Operation' and 'Operands' whereas operand data are saved in different register in every cycle. The Figure 8.1(b) and Figure 8.1(c) show consecutive shift operations and store operations with different 'Operand' while keeping same 'Other Operations' in every cycle. It means that the context words show consecutively same part and they are repetitively read from configuration cache without changing values.

8.2.2.3   Redundancy Ratio

For statistical evaluation of redundant context words, we selected 32-bit context architecture of the base architecture (Figure 4.3) and mapped several kernels onto its PE array in order to maximize the utilization of the context fields. Figure 8.2 shows the results for various benchmark kernels and critical loops in real applications. Each kernel shows three cases of redundancy ratios—'NOP', 'Consecutively Same' and Total. Total redundancy ratio varies from 31% to 75%.

8.3   Dynamic Context Management

If the configuration cache does not perform read/write-operation for redundant part of context words, it is possible to reduce power consumption in configuration cache. That way, one can achieve low-power implementation of CGRA without performance degradation while managing context words in both cases at transfer time and runtime: one case is no read/write-operation

Cycle time	Context Word		
	ALU Operation	**Operands**	**Other Operations**
1			R0 <= ALU_OUT
2	ALU_OUT <= A	A<= Data bus	R1 <= ALU_OUT
3			R2 <= ALU_OUT
4			R3 <= ALU_OUT

(a)

Cycle time	Context Word		
	ALU Operation	**Operands**	**Other Operations**
1	ALU_OUT <= AXB	A <= T, B<= L	
2	ALU_OUT <= A+B	A <=BT, B <= R	SHIFT(ALU_OUT)
3	ALU_OUT <= A-B	A <= R1, B<= R2	
4	ALU_OUT <= A+B	A <= T	

(b)

Cycle time	Context Word		
	ALU Operation	**Operands**	**Other Operations**
1		A <= R0	
2	ALU_OUT <= A	A <= R1	Data bus <= ALU_OUT
3		A <= R2	
4		A <= R3	

(c)

FIGURE 8.1: Consecutively same part in context words: (a) consecutive load operations; (b) consecutive shift operations; (c) consecutive store operations (R0 ∼ R3: registers of register file, T, L, R and BT: output from top PE, left PE, right PE and bottom PE).

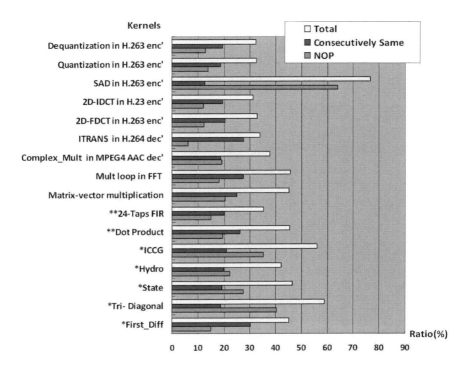

FIGURE 8.2: Redundancy ratio of context words: *Livermore loops benchmark [6], **DSPstone [34] Consecutively Same (%) = 100 (consecutively same part [bits]/total context words [bits]), NOP (%) = 100 (NOP context words [bits] / total context words [bits]), total (%) = NOP + consecutively same.

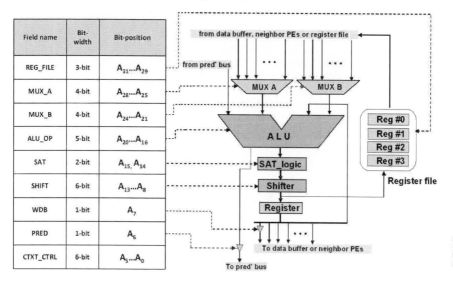

Field name	Bit-width	Bit-position
REG_FILE	3-bit	$A_{31}...A_{29}$
MUX_A	4-bit	$A_{28}...A_{25}$
MUX_B	4-bit	$A_{24}...A_{21}$
ALU_OP	5-bit	$A_{20}...A_{16}$
SAT	2-bit	A_{15}, A_{14}
SHIFT	6-bit	$A_{13}...A_{8}$
WDB	1-bit	A_{7}
PRED	1-bit	A_{6}
CTXT_CTRL	6-bit	$A_{5}...A_{0}$

FIGURE 8.3: An example of PE and context architecture.

for NOP and another case is one read/write-operation for consecutively same part in context words. In order to support such a dynamic context management, we propose a new configuration cache structure and efficient control mechanism in this chapter.

8.3.1 Context Partitioning

Context partitioning is to split context architecture into two parts feasible to dynamic context management. As mentioned in subsection 8.2.2.2, the context words show consecutively same part and they are repetitively read from configuration cache without changing values. Therefore, if a CE is divided into two parts (CE#1 and CE#2) by context partitioning, one part of CE including continuously same part can be disabled for power saving while keeping consecutive read/write-operation of another part of CE. The partitioning starts from grouping context field for ALU operation and some context fields dependent to ALU operation. This is because ALU has the most dependency with other component and they are highly probable to be consecutively changed or un-changed together. Therefore context partitioning positions such fields on one part of context architecture and other fields on another part of context architecture. We have defined generic PE structure and 32-bit context architecture like Figure 8.3 as an example to illustrate context partitioning. It can support the kernels in Figure 8.2. It is similar to the representative CGRAs such as MorphoSys [75], REMARC [61], ADRES [21,82] or PACT_XPP [11]. Bit-width and initial bit-position of each field are shown in Figure 8.3. It supports various arithmetic and logical operations (ALU_OP) with two operands

MSB								LSB
$A_{31}...A_{27}$	$A_{26}...A_{23}$	$A_{22}...A_{19}$	A_{18}	A_{17}, A_{16}	A_{15}	$A_{14}...A_{9}$	$A_{8}...A_{6}$	$A_{5}...A_{0}$
ALU_OP	MUX_A	MUX_B	PRED	SAT	WDB	SHIFT	REG_FILE	CTXT_CTRL
ALU_OP and ALU_OP-dependent Fields : 14-bit				ALU_OP-independent Fields : 18-bit				

FIGURE 8.4: Context partitioning.

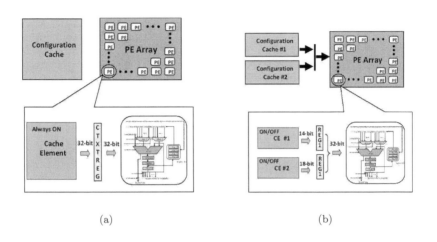

(a)   (b)

FIGURE 8.5: Comparison between general CE and proposed CE: (a) general CE; (b) proposed CE.

(MUX_A and MUX_B), predicated execution (PRED), arithmetic saturation (SAT_logic), shift operation (SHIFT) and saving temporal data with register file (REG_FILE). Figure 8.4 shows context partitioning of Figure 8.3. Field 'ALU_OP' and the fields dependent to 'ALU_OP' are positioned to the part near to MSB and other fields are positioned near to LSB.

After context partitioning, we can know the bit-widths of CE#1 and CE#2 and context register can be also split into two parts with same bit-widths. Figure 8.5 shows comparison between general CE and proposed CE. The proposed CE is composed of CE#1 (14-bit) and CE#2 (18-bit) whereas the general CE is a unified one (32-bit). In subsection 8.3.2 and 8.3.3, we describe more detailed control mechanism for dynamic context management based on the proposed CE structure.

8.3.2   Context Management at Transfer Time

Context management at transfer time is to remove redundant cache-write operations by using additional hardware detecting redundancy of context

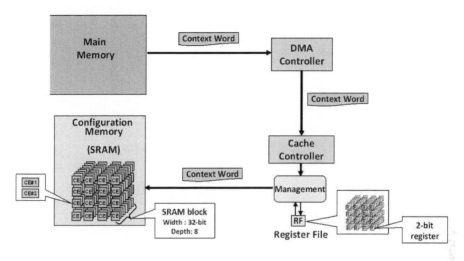

FIGURE 8.6: Context management when context words are transferred.

words. Figure 8.6 shows transfer flow of context words from main memory to configuration cache in the case of 4x4 CEs. For checking the redundancy, hardware block of 'Management' is added to general cache controller. 'Management' block checks transferred context words whether it has redundancy or not. Then it controls cache-write operation as Algorithm 4. In addition, Figure 8.6 shows register file connected with 'Management' block—it has same address-ability as CE but bit-width is 2. The register file stores 2-bit redundancy information—the saved information in register file is used for context management at run time.

Algorithm 4 shows this management process for a CE. Before we explain this management in detail, we introduce notations we use in Algorithm 4.

- *cw*: bit-width of context word

- *w*: bit-width of field group (ALU_OP and ALU_OP-dependent fields)

- *cur_ctxt*: context word currently transferred to configuration cache

- *prev_ctxt*: context word previously transferred to configuration cache

- *ctxt_addr*: address of current context word in configuration cache

- *reg_file*: register file, CE#1 and CE#2: Cache Element

- *out_ctxt*: context word currently provided to context register

- *cs1andcs2*: chip select signal of CE#1 and CE#2

The algorithm starts with checking whether current context word is NOP or not (L2). If the context word is NOP, 2-bit information ("01") is stored in

---

Algorithm 4   Context Management at Transfer Time

L1   **begin**

L2   **if** *cur_ctxt* = NOP **then**

L3   *reg_file*[*ctxt_addr*] ← "01"

L4   *cs1* ← '0', *cs2* ← '0'

L5   **else if** *cur_ctxt*[*cw*-1, *cw*-*w*+1] = *prev_ctxt*[*cw*-1, *cw*-*w*+1] **then**

L6   *reg_file*[*ctxt_addr*] ← "10"

L7   *cs1* ← '0', *cs2* ← '1'

L8   *CE#2*[*ctxt_addr*] ← *cur_ctxt*[*cw*-*w*, 0]

L9   **else if** *cur_ctxt*[*cw*-*w*, 0] = *prev_ctxt*[*cw*-*w*, 0] **then**

L10   *reg_file*[*ctxt_addr*] ← "11"

L11   *cs1* ← '1', *cs2* ← '0'

L12   *CE#1*[*ctxt_addr*] ← *cur_ctxt*[*cw*-1, *cw*-*w*+1]

L13   **else**

L14   *cs1* ← '1', *cs2* ← '1'

L15   *CE#1*[*ctxt_addr*] ← *cur_ctxt*[*cw*-1, *cw*-*w*+1]

L16   *CE#2*[*ctxt_addr*] ← *cur_ctxt*[*cw*-*w*, 0]

L17   **end if**

L18   *prev_ctxt* ← *cur_ctxt*

L19   **end**

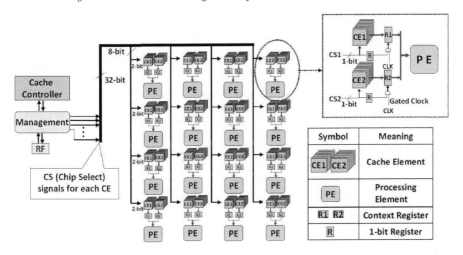

FIGURE 8.7: Context management at run time.

register file and both CE#1 and CE#2 are disabled (L4). If it's not NOP, next process is to check whether the upper part (near to MSB) of context word is consecutively identical to one of previous context word. If it is the same part as the previous one, information ("10") is stored in the register file (L6) and only CE#2 is enabled (L7) for cache write-operation (L8). Checking the lower part (near to LSB) of current context word (L9 ∼ L12) shows the same manner as previous process but CE#1 is enabled instead of CE#2. Finally, if current context word does not correspond to any case of previous checking processes, both CE#1 and CE#2 are enabled (L14) and full context word is stored in configuration cache (L15, L16). Finally, previous context word is updated by current context word (L18).

8.3.3   Context Management at Run Time

Context management at run time is to remove redundant cache-read operations by checking redundancy information stored in the register file. Figure 8.7 shows structure between configuration cache and PE array for the context management. The hardware block of 'Management' controls all of CEs and a context register between a CE and a PE is implemented by a gated clock using chip select signals (CS1 and CS2). Gated clock implementation is to configure PE with fixed output of the context register caused by non-oscillated clock. Therefore, PEs can be configured without cache-read operation in the case of consecutively same context words.

Algorithm 5 shows this management process for a CE. The defined notations in Algorithm 4 are used in Algorithm 5. The algorithm starts with checking whether the information (stored in the register file) identified by current address is NOP or not (L2). If the information is NOP ("01"), both

Algorithm 5	Context Management at Run Time
L1	**begin**
L2	**if** *reg_file*[*ctxt_addr*] = "01" **then**
L3	*cs*1 ← '0', *cs*2 ← '0'
L4	**else if** *reg_file*[*ctxt_addr*] = "10" **then**
L5	*cs*1 ← '0', *cs*2 ← '1'
L6	*out_ctxt*[*cw-w*, 0] ← *CE#*2[*ctxt_addr*]
L7	**else if** *reg_file*[*ctxt_addr*] = "11" **then**
L8	*cs*1 ← '1', *cs*2 ← '0'
L9	*out_ctxt*[*cw-*1, *cw-w+*1] ← *CE#*1[*ctxt_addr*]
L10	**else**
L11	*cs*1 ← '1', *cs*2 ← '1'
L12	*out_ctxt*[*cw-*1, *cw-w+*1] ← *CE#*1[*ctxt_addr*]
L13	*out_ctxt*[*cw-w*, 0] ← *CE#*2[*ctxt_addr*]
L14	**end if**
L15	**end**

CE#1 and CE#2 are disabled (L3). If it's not NOP, next process is to check whether the information corresponds to the case ("10") of consecutively same part (near to MSB) or not (L4). If it is "10", only CE#2 is enabled (L5) for cache read-operation (L6). Next process is to check whether the information corresponds to the case ("10") of consecutively same part (near to MSB) or not (L7). It shows the same manner as previous process but CE#1 is enabled for read-operation instead of CE#2. Finally, if the information does not correspond to any case of previous checking processes, both CE#1 and CE#2 are enabled (L11) and a full context word is read from configuration cache (L12, L13).

8.4   Experiments

8.4.1   Experimental Setup

For quantitative evaluation, we have designed two CGRAs based on the 8x5 reconfigurable array at RT-level with VHDL—one is conventional base CGRA and the other is the proposed CGRA supporting dynamic context management. The architectures have been synthesized using Design Compiler [18] with 0.18 μm technology. We have used SRAM Macro Cell library for the frame buffer and configuration cache. ModelSim [17] and PrimePower [18] tools have been used for gate-level simulation and power estimation. To obtain the power consumption data, we have used the kernels (Figure 8.2)

TABLE 8.1:   Area overhead by dynamic context management

| Component | Area Cost (gate equivalent) | | Overhead (%) |
	Base	Proposed	
Configuration Cache	150012	162538	8.35
Entire RAA	942742	955268	1.33

Base: base architecture, Proposed: proposed architecture,
Overhead(%): {(Proposed/Base) - 1}×100

for simulation with operation frequency of 100 MHz and typical case of 1.8 V
Vdd and 27°C.

8.4.2   Results

8.4.2.1   Area Cost Evaluation

Table 8.1 shows the synthesis results from Design Compiler [18] of proposed
architecture and base architecture. It shows that area cost of new configuration
cache including cache control unit, hardware block of "Management" and
register file increased by 8.35% but the overall area-overhead is only 1.33%.
Thus, the new configuration cache structure can support dynamic context
management with negligible overheads.

8.4.2.2   Power Evaluation

To demonstrate the effectiveness of the proposed approach, we have applied several kernels in Figure 8.2 to the proposed and base architectures.
These kernels were executed with 100 iterations. Table 8.2 shows power evaluation of configuration cache for two cases—read-operation and write-operation.
The power consumptions of write-operations are less than the cases of read-
operations. This is because a CE performs write-operation at transfer time
whereas all of CEs perform read-operation at run time. Compared to the
base architecture, it has shown to save up to 68.89%/69.85% of the power
in write/read-operation. Five kernels (ITRANS, 2D-FDCT, 2D-IDCT, Quant
and Dequant) show less reduction in power compared to other kernels. This
is because they show less redundancy ratios of context words compared with
other kernels—Figure 8.2 shows that the redundancy ratios of these kernels
are in the range of 31.22% ∼ 33.79%. Average power reduction ratios in write-
operation and read-operation are 38.24% and 38.15%.

8.4.2.3   Performance Evaluation

The synthesis results show that the critical path delay of the proposed
architecture is same as the base model i.e. 8.96 ns. It indicates the dynamic
context management does not cause performance degradation in terms of the
critical path delay. In addition, the execution cycle count of each kernel on pro-

TABLE 8.2:   Power reduction ratio by dynamic context management

Kernels	Configuration Cache Power (mW)				Reduction Ratio (%)	
	Write-operation		Read-operation		Write	Read
	Base	Proposed	Base	Proposed		
Tri- Diagonal	14.98	6.89	171.77	79.03	54.00	53.99
First_Diff	13.34	8.25	174.18	104.51	38.12	40.00
State	15.23	9.37	161.23	93.87	38.45	41.78
Hydro	11.22	7.17	148.23	96.14	36.14	35.14
ICCG	15.39	7.56	205.80	103.35	50.87	49.78
Dot Product	12.11	7.28	117.84	72.51	39.88	38.47
24-Taps FIR	19.20	11.63	227.56	138.90	39.41	38.96
MVM	14.23	8.68	227.57	138.54	38.99	39.12
Mult in FFT	12.12	7.62	175.48	105.88	37.14	39.66
Comlex Mult	11.57	7.86	180.63	123.59	32.12	31.58
ITRANS	14.22	10.17	204.85	148.64	28.47	27.44
2D-FDCT	16.23	11.69	190.03	140.30	27.96	26.17
2D-IDCT	17.34	13.16	188.47	139.88	24.13	25.78
SAD	14.30	4.45	185.30	55.87	68.89	69.85
Quant	12.12	8.73	185.23	134.94	27.99	27.15
Dequant	15.33	11.05	187.78	137.10	27.89	26.99
Average					38.24	38.15

Base: base architecture, Proposed: proposed architecture
Reduced: $\{1\text{-}(\text{Proposed}/\text{Base})\} \times 100$
Write/Read: reduction ratio in the case of write/read operation

posed architecture does not vary from the base architecture because the functionality of proposed architecture is same as the base model. It also indicates the dynamic context management does not cause performance degradation in terms of the execution cycle count.

8.5 Summary

In this chapter we describe the dynamic context management for low-power CGRA and the configuration cache structure supporting this technique. The proposed management method can be used to achieve power savings in configuration cache while maintaining performance same as general CGRA. Experiments show that the approach saves much power compared to conventional base model with negligible area overhead. The power reduction ratio is 38.24%/38.15% in write/read operation of configuration cache.

Chapter 9

Cost-Effective Array Fabric

9.1 Introduction

CGRA has higher performance level than general purpose processor and wider applicability than ASIC. However, the deployment of CGRA is prohibitive due to its significant area and power consumption. This is due to the fact that CGRA is composed of several memory components and the array of many processing elements including ALU, multiplier and divider, etc. Especially, processing element (PE) array occupies most of the area and consumes most of the power in the system to support flexibility and high performance. Therefore, reducing area and power consumption in the PE array has been a serious concern for the adoption of CGRA. In this chapter, we introduce a domain-specific array fabric design space exploration method to generate a cost-effective reconfigurable array structure [42]. The exploration flow efficiently rearranges PEs with reducing array size and change interconnection scheme to achieve much reduction in power and area while maintaining the same performance as the original architecture. In addition, the proposed array fabric splits the computational resources into two groups (primitive resources and critical resources). Critical resources can be area-critical and/or delay-critical. Primitive resources are replicated for each processing element of the reconfigurable array, whereas area-critical resources are shared among multiple basic PEs in order to reduce more area of CGRA. Delay-critical resources can be pipelined to curtail the overall critical path so as to increase the system clock frequency. Experimental results show that for multimedia applications, the proposed approach reduces area by up to 36.75%, execution time by up to 42.86% and power by up to 35.45.% when compared with the base CGRA architecture.

9.2 Preliminary

In this section, we present preliminary concepts of our cost-effective design [39]. They come from the characteristics of loop pipelining based on MIMD-style execution model. Then we propose two techniques to make an RAA cost-effective in terms of area and delay. One is resource sharing and the other is resource pipelining.

9.2.1   Resource Sharing

Figure 9.1 shows the snapshot taken at the 5th cycle of execution of the previous example shown in Figure 6.3 for three cases: (a) SIMD and two cases of loop pipelining, (b) temporal mapping and (c) spatial mapping. The operations in the 5th cycle for (a), (b) and (c) include multiplication and therefore the multipliers in the PE array are to be used. In the case of SIMD, all PEs perform multiplication requiring all of them to have multipliers, thereby increasing the area cost of the PE array. However, in the case of temporal mapping, only PEs in the 1st column and the 4th column perform multiplication while PEs in the 2nd and 3rd columns perform addition. In the spatial mapping, only PEs in the 1st and 2nd columns perform multiplication. As can be observed, in the temporal mapping and spatial mapping, there is no need for all PEs to have the same functional resources at the same time. This allows the PEs in the same column or in the same row to share area-critical resources. Figure 9.2 shows four PEs in a row sharing two multipliers at the 5th cycle in temporal mapping and spatial mapping. We depict only the connections related to resource sharing.

Figure 9.3 depicts the detailed connections for multiplier sharing. The two n-bit operands of a PE are connected to the bus switch. The dynamic mapping of a multiplier to a PE is determined at compile time and the information is encoded into the configuration word. At run-time, the mapping control signal from the configuration word is fed to the bus switch and the bus switch decides where to route the operands. After the multiplication, the 2n-bit output is transferred from the multiplier to the original issuing PE via the bus switch.

9.2.2   Resource Pipelining

If there is a critical functional resource with long latency in a PE, the functional resource can be pipelined to curtail the critical path. Resource pipelining has clear advantage in loop pipelining execution because heterogeneous functional units with different delays can run at the same time. In the traditional design (Figure 9.4(a)), the latency of a PE is fixed but in our pipelined PE design (Figure 9.4(b)), we allow multi-cycle operations and so the latency can vary depending on the operation. This helps increase the system clock frequency.

If a critical functional resource such as a multiplier has both large area and long latency, the resource sharing and resource pipelining can be applied at the same time in such a way that the shared resource executes multiple operations at the same time in different pipeline stages. With this technique, the conditions for resource sharing are relaxed and so the critical resources are

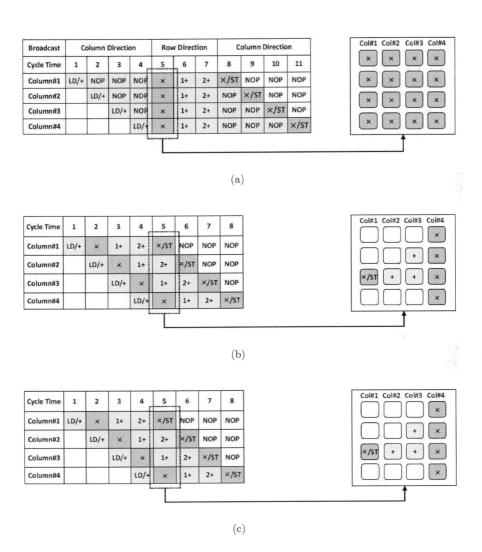

FIGURE 9.1: Snapshots of three mappings: (a) SIMD; (b) temporal mapping; (c) spatial mapping.

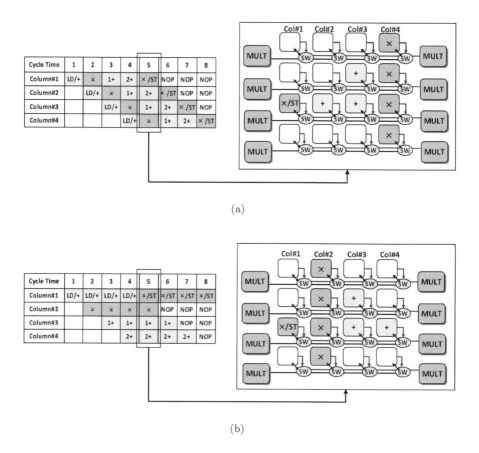

Cycle Time	1	2	3	4	5	6	7	8
Column#1	LD/+	×	1+	2+	× /ST	NOP	NOP	NOP
Column#2		LD/+	×	1+	2+	× /ST	NOP	NOP
Column#3			LD/+	×	1+	2+	× /ST	NOP
Column#4				LD/+	×	1+	2+	× /ST

(a)

Cycle Time	1	2	3	4	5	6	7	8
Column#1	LD/+	LD/+	LD/+	LD/+	× /ST	× /ST	× /ST	× /ST
Column#2		×	×	×	×	NOP	NOP	NOP
Column#3			1+	1+	1+	1+	NOP	NOP
Column#4				2+	2+	2+	2+	NOP

(b)

FIGURE 9.2: Eight multipliers shared by sixteen PEs: (a) temporal mapping; (b) spatial mapping.

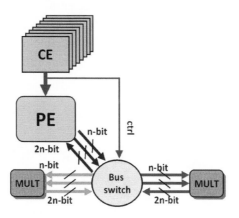

FIGURE 9.3: The connection between a PE and shared multipliers.

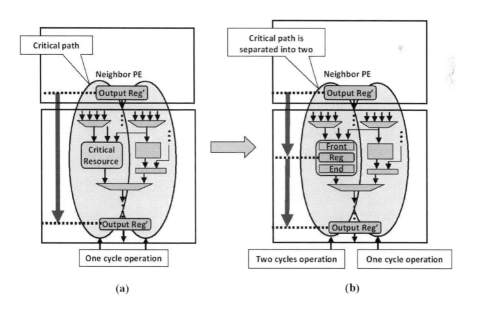

FIGURE 9.4: Critical paths: (a) general PE; (b) pipelined PE.

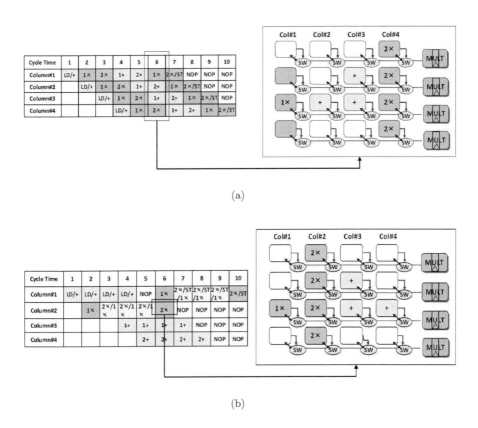

(a)

(b)

FIGURE 9.5: Loop pipelining with pipelined multipliers: (a) temporal mapping; (b) spatial mapping (1: first pipeline stage on multiplication, 2: second pipeline stage on multiplication).

utilized more efficiently. Figure 9.5 shows this situation. Through the pipelining, we can reduce the number of multipliers from 8 to 4 to perform the execution without any stall. This is because two PEs sharing one pipelined multiplier can perform two multiplications at the same time using different pipeline stages.

9.3   Cost-Effective Reconfigurable Array Fabric

In this section, we propose an array fabric design space exploration method to generate a cost-effective reconfigurable array structure in terms of area and power. It is mainly motivated by the characteristics of typical computation-intensive and data-parallel applications.

9.3.1   Motivation

9.3.1.1   Characteristics of Computation-Intensive and Data-Parallel Applications

Most of the CGRAs have been designed to satisfy the performance requirement of a range of applications in a particular domain. Especially, they have been designed for applications that exhibit computation-intensive and data-parallel characteristics. Common examples for such applications are digital signal processing (DSP) applications like audio signal processing, image processing, video signal processing, speech signal processing, speech recognition, and digital communications. Such applications have many sub-tasks such as trigonometric functions, filters and matrix/vector operations that can be mapped onto coarse-grained reconfigurable array. We have classified such subtasks into four types as shown by the data flow graphs in Figure 9.6. Type (a) shows merge operation in which outputs from multiple operations in the previous stage are used as inputs to an operation in the next stage. Type (b) shows butterfly operation where output data from multiple operations in the previous stage are fed as input data to the same number of next stage operations. Finally, type (c) and (d) show the combinations of (a) and (b).

9.3.1.2   Redundancy in Conventional Array Fabric

Most coarse-grained reconfigurable arrays arrange their processing elements (PEs) in a square or rectangular 2-D array with rich set of horizontal and vertical connections for effective exploitation of parallelism. However, such square/rectangular array structures have many redundant or unutilized PEs during the executions of applications on them. Figure 9.7 shows an example of three types of data flow (Figure 9.6(a), Figure 9.6(c), Figure 9.6(d)) mapped onto 8x8 square reconfigurable arrays in the two cases of loop pipelining—

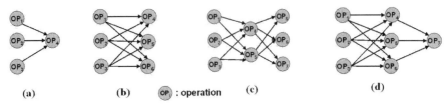

FIGURE 9.6: Subtask classification.

FIGURE 9.7: Data flow on square reconfigurable array.

temporal mapping and spatial mapping. The upper part of Figure 9.7 shows the scheduling for a column of PEs based on temporal mapping and also shows how the utilization of the PEs changes for the 8 cycles of schedule. As can be seen from the figure, some PEs have very low utilization. The lower part of Figure 9.7 shows the spatial mapping of the 8x8 array, where some PEs are not used at all. All the three types of implementations show lots of redundant PEs that are not used.

From these observations, we see that the existing square/rectangular array fabric cannot efficiently utilize the PEs in the array and therefore waste large area and power. In order to overcome such wastages in square/rectangular array fabric, we propose a new cost-effective array fabric in the next subsection.

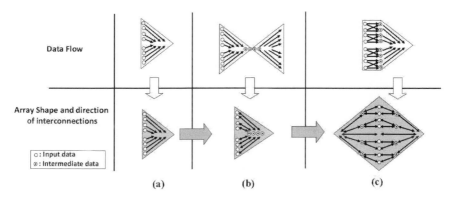

FIGURE 9.8: Data flow-oriented array structure derived from three types of data flow.

9.3.2 New Cost-Effective Data Flow-Oriented Array Structure

9.3.2.1 Derivation of Data Flow-Oriented Array Structure

To reduce the redundancy in the conventional square/rectangular array, first of all, we can consider a specific array shape that fits well with the applications' common data flows. Figure 9.8 shows such a data flow-oriented array structure derived from three types of data flow. In Figure 9.8(a), a triangular-shaped array and uni-directional interconnections among PEs can be derived from the first data flow (merge). Then the interconnections can be made bi-directional to support the merge-butterfly data flow as shown in Figure 9.8(b). Finally, in Figure 9.8(c), the entire array becomes a diamond-shaped structure to reflect the butterfly-merge data flow. In this case, the butterfly operations are spatially spread on both sides of the array. Then intermediate data merge takes place at the end of both sides or they can merge at the center of the array.

To represent how the data-flow oriented array structure can efficiently utilize PEs, we examine the difference between the conventional square-shaped array and the proposed data flow-oriented array with a simple example. We assume a diamond-shaped reconfigurable array composed of 12 PEs as shown in Figure 9.9(a)—this is a counterpart of the 4x4 PE array shown in Figure 6.1. In addition, we assume a Frame Buffer similar to the one in Figure 6.1(b) is connected to the array, where the PEs in each row of the array share two read buses and the PEs in two neighboring rows share one write bus as shown in Figure 9.9(b). The array has nearest neighbor and global bus interconnections in diagonal and horizontal directions as shown in Figure 9.9(c) and Figure 9.9(d).

Consider mapping of Eq. (6.1) in Chapter 6 with N = 4 on the proposed array in the same way as we did for the 4x4 square-shaped PE array in Chapter

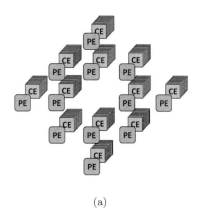

(a)

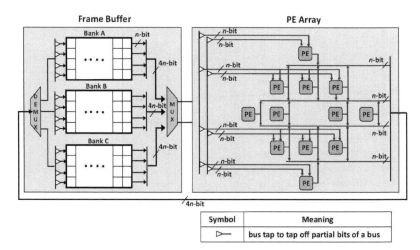

(b)

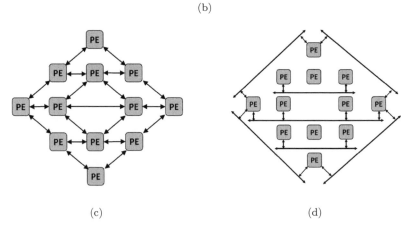

(c)   (d)

FIGURE 9.9: An example of data flow-oriented array: (a) distributed cache structure; (b) frame buffer and data bus; (c) nearest neighbor interconnection; (d) global bus interconnection.

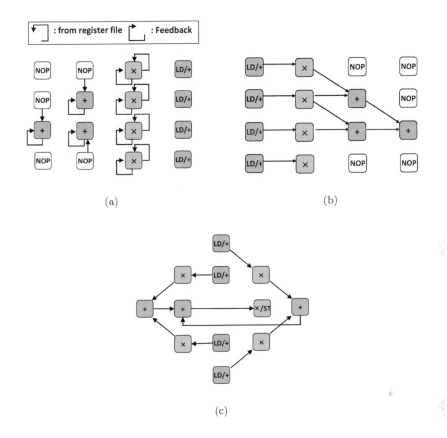

FIGURE 9.10: Snapshots showing the maximum utilization of PEs: (a) temporal mapping on 4x4 PE array; (b) spatial mapping on the 4x4 PE array; (c) spatial mapping on the data flow-oriented PE array.

6. Figure 9.10 shows the snapshots taken at the time of maximum utilization of PEs for three cases: (a) temporal mapping on the square-shaped array (4x4 PEs), (b) spatial mapping on the square-shaped array (4x4 PEs) and (c) spatial mapping on the data flow-oriented array (12 PEs). In the case of (a) and (b), five PEs are not used because the merging addition does not fit well with the square-shape. However, in the case of (c), the array efficiently utilizes all of the PEs without delayed operation. As can be observed, this example shows that the proposed array structure can avoid the area and power wastages of the square-shaped array without performance degradation.

9.3.2.2   Mitigation of Spatial Limitation in the Proposed Array Structure

As shown in Figure 9.10(c), we spread the operations in the data flows (mostly loop bodies) over the array space, instead of spreading the operations over time for each column to implement temporal loop pipelining as shown in Figure 9.10(a). This implies that spatial loop pipelining is most suitable to the new array fabric. However, as mentioned in Chapter 6 (see subsection 6.2.2), spatial mapping is not feasible for complex loops because of two reasons. One is that a large loop body may not fit in the limited reconfigurable array and the other is that data dependencies between the operations typically require allocating lots of interconnect resources. In order to mitigate such a limitation, the new array fabric should have rich interconnections to provide more flexible and multi-directional data communication and the PEs should be arranged in such a way to utilize such an interconnection structure efficiently. As a solution to this problem, we propose a design flow that generates a data flow-oriented array structure by determining the topology of PEs and their interconnections.

9.3.3   Data Flow-Oriented Array Design Flow

The generation of a data flow-oriented array starts from a square-shaped array fabric, considering that the original square-shaped array fabric is very well designed. We generate the new data flow-oriented array such that it can efficiently implement any application that can be implemented on the square-shaped array fabric. In the example of Figure 9.9(a), since the data flow has a diamond-shape, we can generate a diamond-shaped array with less number of PEs but without any performance degradation, which is just like garment cutting. Since we want to cover the applications that can be implemented through temporal mapping on the square-shaped array fabric as well, we do not just cut the fabric but we compose the new array by transforming the temporal interconnection structure to a spatial interconnection structure. In the temporal mapping, each loop iteration of an application kernel (critical loop) is mapped onto a column of the square-shaped array. Therefore, it is good enough to analyze the interconnection fabric only within a column to derive the new array structure.

Figure 9.11 shows the entire design flow. This flow starts from analysis of *intra-half column* and *inter-half column connectivity* of general square array fabric. *Intra-half column connectivity* means nearest neighbor or hopping interconnection between PEs in a half column and *inter-half column connectivity* means pair-wise interconnection or global bus between PEs—one PE in a half column and another PE in the other half column. New array fabric is partially derived by analyzing intra-half column connectivity of square fabric in Phase I. Then Phase II elaborates new array fabric by analyzing inter-half column connectivity of square fabric. Finally, the connectivity of new array fabric is enhanced by adding vertical and horizontal global bus. In the remain-

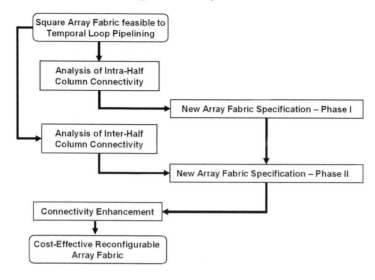

FIGURE 9.11: Overall design flow.

der of this subsection—from 9.3.3.1 through 9.3.3.4 below—we describe more detailed process for each stage of the entire exploration flow.

9.3.3.1 Input Reconfigurable Array Fabric

The 8x8 array given in Figure 4.4 is used for the input array fabric to illustrate the proposed design flow.

9.3.3.2 New Array Fabric Specification - Phase I

In this phase, an initial version of the new array fabric is constructed by analyzing intra-half column connectivity of the input square array. Algorithm 6 shows this procedure. Before we explain the procedure in detail, we describe the notations used in it.

- (L1) *basecolumn* denotes a half column in the n x n reconfigurable array.

- (L3) *new_array_space* denotes 2-dimensional space of the constructed reconfigurable array.

- (L5) *source_column_group* denotes a group of PEs composed of one or two columns in the *new_array_space*. It is used as a source for deriving the new array fabric.

- (L12) |*source_column_group*| denotes the number of PEs in *source_column_group*.

- (L6) CHECK_INTERCONNECT is a function to identify nearest neighbor or hopping interconnections of PEs in *source_column_group* by

---

Algorithm 6 New Array Fabric Specification – Phase I

L1	*base* ← a half column of *n* x *n* reconfigurable array
L2	*m* ← number of memory-read buses of *n* x *n* reconfigurable array
L3	*new_array_space* ← O
L4	**begin**
L5	*source_column_group* ← Add a column composed of n/2 PEs in *new_array_space*
L6	**while** CHECK_INTERCONNECT(*source_column_group*, *base*)
L7	**do**
L8	LOC_TRI(*source_column_group*)
L9	**end do**
L10	*source_column_group* ← O
L11	*source_column_group* ← next two columns on the both sides in *new_array_space*
L12	**if** |*source-column_group*| > 2 **then**
L13	**goto** L6
L14	**end if**
L15	Add nearest-neighbor interconnections
L16	Add *m* memory-read buses
L17	Connect the read buses with the added PEs in the same row
L18	Copy the constructed fabric on vertically symmetric position
L19	**end**

---

analyzing the base column. If there is such an interconnection that has not been processed yet, then it returns true.

- (L8) LOC_TRI is a function that implements the *localtriangulation* method, which adds PEs, assigns them new positions in *new_array_space*, and connects them with the PEs already existing in the *source_column_group*.

The algorithm starts with the initialization step (L1 ∼ L3). Then a half column is added into *new_array_space*, which is the initial *source_column_group* (L5). The next process is to check the nearest neighbor or hopping connectivity between two PEs (L6) in the same column included in *source_column_group*. This checking process (L6) continues until no more interconnection is found. The first checking process is performed by simply identifying interconnections of the base column. If an interconnection is found, two PEs are added into *new_array_space* and their interconnections and positions are assigned by *local triangulation method* (L8). This method is to reflect intra-half column connectivity with making the data flow-oriented array structure as shown in Figure 9.9.

We illustrate the basic concept of local triangulation method in Figure 9.12. Consider a base column including 2 PEs and let the operation fully utilize their interconnections as shown in Figure 9.12(a)—data (A_1 and A_2)

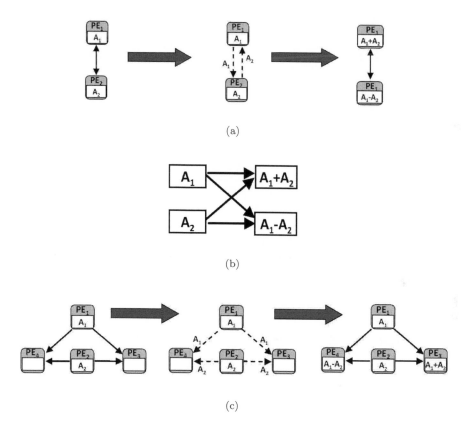

FIGURE 9.12: Basic concept of local triangulation method: (a) an operation fully utilizing interconnections between two PEs; (b) butterfly operation example; (c) butterfly operation executed on a triangle structure including four PEs.

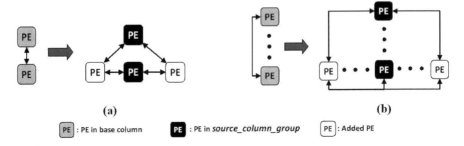

(a)   (b)

PE : PE in base column   PE : PE in *source_column_group*   PE : Added PE

FIGURE 9.13: Local triangulation method.

saved in two PEs (PE1 and PE2) are exchanged with each other through the bidirectional interconnection, and then addition and subtraction are performed in PE1 and PE2. This is a kind of butterfly operation and Figure 9.12(b) shows an equivalent data flow graph for the butterfly operation. If we consider a triangular structure composed of four PEs reflecting the shape of the data flow graph, the example can be mapped on the PEs as shown in Figure 9.12(c)—two PEs (PE3 and PE4) on both sides receive the data (A_1 and A_2) from the PEs (PE1 and PE2), and then addition and subtraction are performed in PE3 and PE4. In such a manner, *local triangulation* method is to make a data flow-oriented array structure reflecting the intra-half column connectivity.

Figure 9.13 shows two cases of the method. In the first case (a), two PEs in the base column have nearest-neighbor interconnection, which means maximum two PEs can be used for butterfly operation. Therefore, *local triangulation* method adds two PEs into *new_array_space* and assigns each PE the nearest-neighbor position on each side of the source column and the positions are vertices of a triangle. Then the method assigns nearest-neighbor interconnection between added PEs and the PEs in the *source_column_group*. The second case (b) shows that two PEs in the base column have a bidirectional hopping interconnection. Local triangulation method is also applied to this case with the hopping interconnections instead of the nearest neighbor interconnections for the first case. Even though one-way interconnections are sufficient to perform butterfly operation in two cases of Figure 9.13, the added interconnections are bidirectional. This is because it aims to keep the basic characteristics of the data flow-oriented array structure derived in Figure 9.9.

From the second checking process, preoccupied columns are included in *source_column_group*. Figure 9.14 shows two examples on how to find connectivity on the source columns. In the case of (b), no interconnection between 'PE4' and 'PE6' (or 'PE7' and 'PE9') is added because there is no hopping connectivity between 'PE0' and 'PE2' (or 'PE0 (or PE1)' and 'PE3') in the base column. However, in the case (c), the base column has interconnection between 'PE0' and 'PE2'. Therefore PEs and interconnections between 'PE4' and 'PE6' (or 'PE7' and 'PE9') are added by local triangulation method.

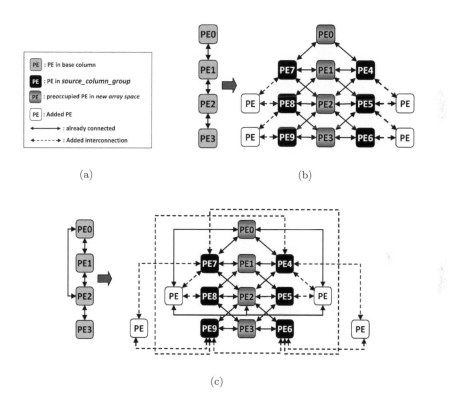

(a)

(b)

(c)

FIGURE 9.14: Interconnection derivation in Phase I: (a) symbols and their meaning; (b) nearest-neighbor; (c) hopping.

After the iteration of adding PEs and interconnections (L5 ∼ L14) is finished, nearest-neighbor interconnections are added between two nearest-neighbor PEs that are not connected with each other (L15). It is to guarantee the minimum data-mobility for data rearrangement. Finally, memory-read buses are added (L16, L17) and the derived array is copied to the vertically symmetric position (L18). Figure 9.15 shows the result of the phase I procedure for the example of 8x8 reconfigurable array as shown in Figure 4.4.

9.3.3.3   New Array Fabric Specification - Phase II

In this phase, new PEs and interconnections are added for reflecting intra-half column connectivity of the input square fabric. Phase II analyzes two kinds of interconnections—pair-wise and global bus. We propose another procedure as Algorithm 7. Before we explain the procedure in detail, we introduce notations we use in the explanation.

- (L5) CHECK_INTERCONNECT is a function to identify bus-connectivity between two PEs in the source column by analyzing the base column.

- (L6) GB_RHT_TRI means *global triangulation* method that is a function used to add global buses and PEs.

The algorithm starts with initialization step (L1, L2). Then central column in *new_array_space* is initial *source_column_group*. Next process is to check the pair-wise or global bus connectivity between two PEs (L5)—two PEs in the same column included in *source_column_group*. If an interconnection is found, global buses and PEs are added in *new_array_space* and their interconnections and positions are assigned by *global triangulation* method (L6). *Global triangulation* method has the same basic concept of *local triangulation* method in that the method is also to make a triangular-shaped array fabric suitable spatial mapping with guaranteeing the maximum inter-half column connectivity of the base column. Figure 9.16 and Figure 9.17 show three cases of *global triangulation* method when the base column has a bidirectional pair-wise interconnection and two global buses. In the first case (a), the bidirectional pair-wise interconnection means maximum two PEs can be used for butterfly operation. Therefore, *global triangulation* method adds two PEs in new array space and assigns each PE the intersection point of two diagonal lines from two PEs in source column. The positions are vertices of a triangle. Then the method assigns four global buses between added PEs and the PEs in the *source_column_group*. Figure 9.16(b) and Figure 9.17(a) show the method when the base column has two global buses. In the case of (b), two diagonal lines from two PEs in source column intersect on already existing PE called 'destination PE'. Therefore, four global buses are added and they connect destination PEs with PEs in the source column. However, in the case of (c), no destination PE exists on intersection point of four diagonal lines. There-

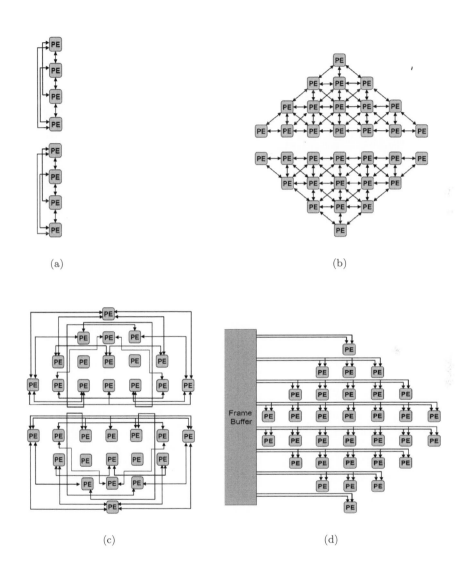

FIGURE 9.15: New array fabric example by Phase I: (a) base column; (b) nearest-neighbor interconnection; (c) hopping interconnection; (d) memory read-buses.

Algorithm 7 New Array Fabric Specification - Phase II

L1   *base* ← a column of n x n reconfigurable array

L2 n ← number of global buses

L3   **begin**

L4   *source_column_group* ← central column in *new_array_space*

L5   **while** CHECK_INTERCONNECT(*source_column_group*, *base*) **do**

L6   GB__TRI (*source_column*)

L7   **end do**

L8   *source _column_group* ← Ø

L9   *source _column_group* ← next two columns on the both sides

                            in *new_array_space*

L10   **if** |*source-column_group*| > 2 **then**

L11   **goto** L5

L12   **end if**

L13   Add nearest-neighbor interconnections

L14   **end**

---

fore, new PEs called global PE (GPE) as well as global buses a/fre added on *new_array_space*.

This checking process (L5) continues until no more connectivity is found. Then nearest-neighbor interconnections are added between two PEs not connected with each other. This is to guarantee the minimum data-mobility for data rearrangement. Figure 9.18 shows the result of the phase II procedure for the example of 8x8 reconfigurable arrays.

9.3.3.4   Connectivity Enhancement

Finally, vertical and horizontal bus can be added to enhance connectivity of new reconfigurable array. This is because new array fabric from phase I and II only has nearest neighbor or hopping interconnection in vertical and horizontal direction whereas it supports sufficient diagonal connectivity. Added horizontal bus is used as memory-write bus connected with frame buffer as well as used for data-transfer between PEs. Figure 9.19 shows the result of the connectivity enhancement for the example of 8x8 reconfigurable array. Each bus is shared by two PEs in both the sides.

9.3.4   Cost-Effective Array Fabric with Resource Sharing and Pipelining

The resource sharing and pipelining mentioned in Section 9.2 can be applied to the proposed new array fabric because the computation model for the proposed fabric is spatial loop pipelining—spatial mapping spreads the entire loop body on the PE array, there is no need for all PEs to have the same functional resources at the same time. Figure 9.20 shows the PEs in the

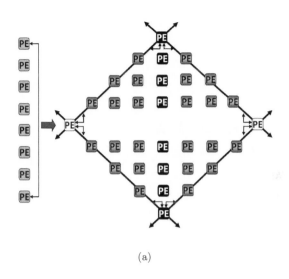

(a)

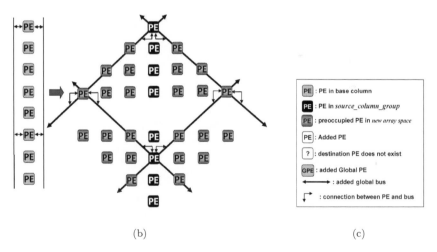

(b)

(c)

FIGURE 9.16: Global triangulation method when n = 2 (L2): (a) when pair-wise interconnection exists in base column; (b) when destination PE exists; (c) symbol definition.

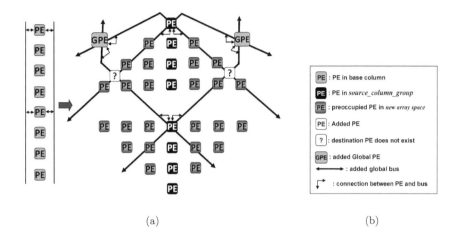

(a)   (b)

FIGURE 9.17: Global triangulation method when n = 2 (L2): (a) when GPE is added; (b) symbol definition.

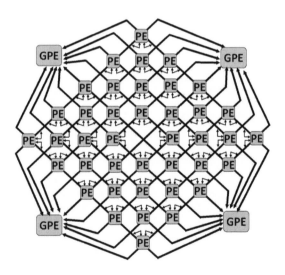

FIGURE 9.18: New array fabric example by Phase II.

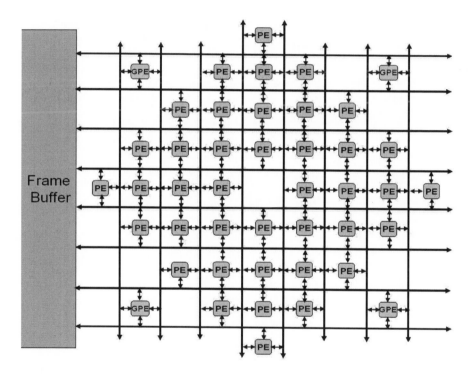

FIGURE 9.19: New array fabric example by connectivity enhancement.

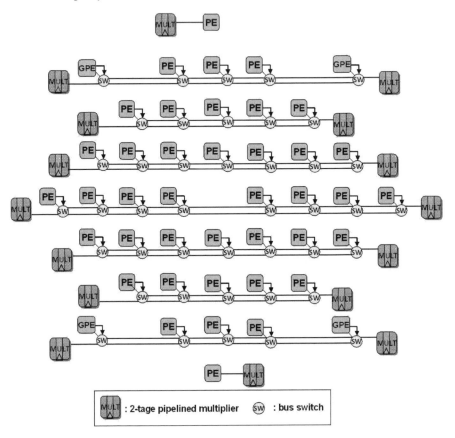

FIGURE 9.20: New array fabric with resource sharing and pipelining.

same row share two pipelined multipliers. Figure 9.21 shows an application
mapping on new array fabric example generated from the exploration flow—
consider N = 8 for the mapping of the computation defined in Eq. (6.1) in
Chapter 6 on the new array fabric as shown in Figure 9.10. Load and addition
operations in PEs are executed on the central column in the first cycle. Then
the next multiplications and summations are spatially spread on both sides
of the array till 6th cycle. Finally, in next two cycles, a PE in the central
row performs multiplication/store operations. The architecture including 16
multipliers supports the mapping example without stall caused by multiplier
lack. In this example, nearest-neighbor interconnections and global buses are
efficiently used for multi-directional data-transfer and the new array has the
same performance (number of execution cycles) compared with the square
array.

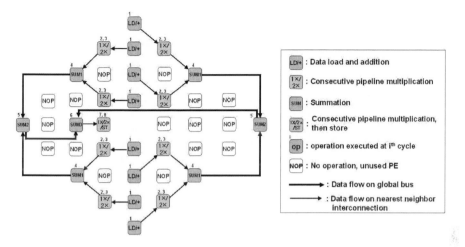

FIGURE 9.21: Mapping example on new array fabric.

9.4 Experiments

9.4.1 Experimental Setup

9.4.1.1 Evaluated Applications

The target application-domain is composed of representative kernels in MPEG-4 AAC decoder, H.263 encoder, H.264 decoder, and 3D-graphics. In addition, to demonstrate the effectiveness of our approaches for benchmark domains, we have applied several kernels of Livermore loops benchmark [6] and DSPstone [34].

9.4.1.2 Hardware Design and Power Estimation

To demonstrate the effectiveness of Resource Sharing and Pipelining (RSP), we have applied RSP techniques to the base RAA (BASE) defined in Chapter 4 (see Section 4.3) and implemented the RSP architecture (RSPA) at the RT-level with VHDL. In Chapter 4 (see Section 4.4), we have confirmed that multiplier is both area-critical and delay-critical resources. Therefore we have taken the multiplier out of the PE design and arranged them to be shared and pipelined resources. From the analysis of our target applications, we have determined the sharing architecture—two pipelined multipliers shared by 8 PEs in each row. Therefore, the RSP architecture including 16 multipliers supports all of the target applications without stall caused by multiplier lack. In addition, we have implemented entire exploration flow in Figure 9.11 with C++. The implemented exploration flow has generated the specification of new reconfigurable array fabric. The base RAA (Chapter 4) has been used for input of the exploration flow. For quantitative evaluation, we have designed two cases of PE array based on the generated specification at the RT-level with VHDL—only new array fabric (NAF) and NAF with RSP (RSP+NAF) - 18 multipliers are shared by PEs in both row and column directions and this architecture also supports all of the target applications without stall caused by multiplier lack. The architectures have been synthesized using Design Compiler [18] with 0.18 μm technology. ModelSim [17] and PrimePower [18] are used for gate-level simulation and power estimation. Simulation has been done for the typical case under the condition of 100 MHz operation frequency, 1.8 V Vdd, and 27°C temperature.

9.4.2   Results

9.4.2.1   Area Evaluation

Table 9.1 shows area cost evaluation for the four cases. In RSPA, the area cost of PE array is reduced by 22.11% because it has less multipliers than BASE. In the case of NAF, the area reduction ratio (25.58%) has relatively increased compared to RSPA. This is because the number of PEs is reduced. Finally, the area reduction ratio (36.75%) in RSP+NAF has also relatively increased compared to NAF because of reduced multipliers. However, the interconnect area of the RSPA (or RSP+NAF) has increased compared to the BASE (or NAF). This is because several buses are added to connect the shared multipliers with PEs.

9.4.2.2   Performance Evaluation

The synthesis results show that RSPA has reduced critical path delay (5.12 ns) compared to BASE (8.96 ns). This is because RSP technique excludes the combinational logic path of the multiplier from the original set of critical paths. The critical path of RSPA and its delay is given by

$$T_{critical\_path} = T_{Multiplexor} + T_{ALU} + T_{Shift\_logic} + T_{others} \qquad (9.1)$$
$$(5.12 \text{ ns} = 0.32 \text{ ns} + 2.22 \text{ ns} + 1.42 \text{ ns} + 1.16 \text{ ns})$$

Table 9.2 shows that BASE and NAF (or RSPA and RSP+NAF) have same critical path delay. It indicates NAF does not cause performance degradation in terms of the critical path delay. In addition, the execution cycle count of each kernel on NAF (or RSP+NAF) does not vary from BASE (or RSPA) because the functionality of NAF is same as the base model. It also indicates NAF does not come by performance degradation in terms of the execution cycle count.

We have applied application kernels to the implemented architectures to obtain the results in Table 9.2. The amount of performance improvement depends on the application. For example, compared to DCT and hydro having multiplication, we achieve much more performance improvement with RSPA

TABLE 9.1: Area reduction ratio by RSPA and NAF

| PE Array Structure | No' of PEs | No' of Multipliers | Gate Equivalent | | | Reduction Ratio (%) |
			Interconnect[a]	Logic[b]	Total[c]	
BASE	64	64	164908	494726	659635	-
RSPA	64	16	170008	343781	513789	22.11
NAF	44	44	156163	334737	490900	25.58
RSP+NAF	44	18	164414	252805	417219	**36.75**

Interconnect[a]: net interconnect area, Logic[b]: total cell area,
Total[c] : Interconnect[a] + Logic[b], Reduction Ratio (%): compared with BASE

TABLE 9.2:   Applications characteristics and performance evaluation

Kernels	Operations[c]	BASE and NAF (8.96 ns)[d]		RSPA(+NAF) (5.12 ns)[d]		
		Cycle count	[e]ET(ns)	Cycle count	[e]ET(ns)	[f]R (%)
[a]First_Diff	sub	15	134.40	15	76.80	42.86
[a]Tri-Diagonal	sub, mult	17	152.32	18	92.16	39.50
[a]State	add, mult	20	179.20	23	117.76	34.29
[a]Hydro	add, mult	15	134.40	19	97.28	27.62
[a]ICCG	sub, mult	18	161.28	19	97.28	39.68
[b]Inner Product	add, mult	21	188.16	22	112.64	40.14
[b]24-Taps FIR	add, mult	20	179.2	21	107.52	40.00
Matrix-vector multiplication	add, mult	19	170.24	20	102.4	39.85
Mult in FFT	add, sub, mult	23	206.08	27	138.24	32.92
Comlex Mult in AAC decoder	add, sub, mult	16	143.36	17	87.04	39.29
ITRANS in H.264 decoder	add, sub, shift	18	161.28	18	92.16	42.86
DCT in H.263 encoder	add, sub, shift, mult	32	286.72	40	204.80	28.57
IDCT in H.263 encoder	add, sub, shift, mult	34	304.64	42	[2]15.04	29.41
SAD in H.263 encoder	add, abs	39	349.44	39	199.68	42.86
Quant in H.263 encoder	add, sub, shift, mult	39	349.44	45	230.40	34.07
Dequant in H.263 encoder	add, sub, shift, mult	41	367.36	57	240.64	34.49

[a]Livermore loop benchmark suite. [b]DSPstone benchmark suite.
[c]Acronym of operations, add: addition, sub: subtraction, shift: bit-shift, mult: multiplication. [d]Critical path delay.
[e]Execution time = cycle×critical path delay(ns).
[f]Reduction ratio of execution time compared with BASE.

TABLE 9.3: Power reduction ratio by RSP+NAF

Kenels	PE Array Structure		
	BASE	RSP+NAF	
	Power (mW)	Power (mW)	[a]R(%)
First_Diff	201.07	129.79	35.45
Tri-Diagonal	190.75	130.89	31.38
State	198.37	138.62	30.12
Hydro	190.86	129.35	32.23
ICCG	164.42	112.92	31.32
Inner Product	200.09	139.30	30.38
24-Taps FIR	174.38	116.40	33.25
Matrix-vector mult'	163.25	113.48	30.49
Mult in FFT	187.68	125.30	33.24
Comlex Mult	222.14	148.55	33.13
ITRANS	198.32	137.89	30.47
DCT	212.25	147.90	30.32
IDCT	208.99	143.58	31.30
SAD	181.22	123.23	32.00
Quant	199.38	137.33	31.12
Dequant	196.97	131.28	33.35
[a]Power reduction ratio compared with BASE			

and RSP+NAF for First_Diff, SAD, and ITRANS which have no multiplication. This is because the clock frequency has been increased by pipelining the multipliers whereas the execution cycle count does not vary from BASE and NAF.

9.4.2.3 Power Evaluation

Table 9.3 shows the comparison of power consumptions between the two reconfigurable arrays: BASE and RSP+NAF. The two arrays have been implemented without any low-power technique to evaluate their power savings. It is shown that compared to BASE, RSP+NAF could save up to 35.45% of the power. It has been possible to reduce power consumption in RSP+NAF by using less number of PEs and multipliers to do the same job compared to the base reconfigurable array. For larger array sizes, the power saving will further increase due to significant reduction in unutilized PEs.

9.5 Summary

In this chapter, we describe a CGRA design space exploration flow optimized for computation-intensive and data parallel applications. First of all, it has been shown that the new data flow-oriented array fabric is derived from a standard square-array using the proposed exploration flow. The exploration flow efficiently rearranges PEs with reducing array size and changes interconnection scheme to save area and power. In addition, we suggest the design scheme which splits the computational resources into two groups (primitive resources and critical resources). Primitive resources are replicated for each processing element of the reconfigurable array, whereas area-critical resources are shared among multiple basic PEs. Delay-critical resources can be pipelined to curtail the overall critical path delay so as to increase the system clock frequency. Experimental results show that the approach saves significant area and power with enhanced performance compared to conventional base model. Implementation of sixteen kernels on the new array structure demonstrates consistent results. The area reduction up to 36.75%, the average performance enhancement of 36.78%, and the average power saving of 31.85% are evident when compared with the conventional array architecture.

Chapter 10

Hierarchical Reconfigurable Computing Arrays

10.1 Introduction

Coarse-grained reconfigurable architecture (CGRA) based embedded system aims at achieving high system performance with sufficient flexibility to map variety of applications. However, significant area and power consumption in the arrays prohibits its competitive advantage to be used as a processing core. In this chapter, we present a computing hierarchy consisting of two reconfigurable computing blocks with two types of communication structure together [45]. In addition, the two computing blocks have shared critical resources. Such a sharing structure provides efficient communication interface between them with reducing overall area. Based on the proposed architecture, optimized computing flows have been implemented according to the varying applications for low power and high performance. Experimental results show that the proposed approach reduces on-chip area by 22%, execution time by up to 72% and reduces power consumption by up to 55% when compared with the conventional CGRA-based architectures.

10.2 Motivation

10.2.1 Limitation of Existing Processor-RAA Communication Structures

A typical coarse-grained reconfigurable architecture consists of a microprocessor, a Reconfigurable Array Architecture (RAA), and their interface. We can consider three types of organizations in connecting RAA to the processor. First, the array can be connected to the processor through a system bus as an 'Attached IP' [11, 19, 35, 46, 70, 75] shown in Figure 4.1(a). In this case, the main benefit of this organization is the ease of constructing such a system using a standard processor without modifying the processor and its compiler. In

TABLE 10.1:   Comparison of the basic coupling types

Coupling type	*Comm' power	**Comm' speed	Performance bottleneck	Application feasibility
Attached IP	high	slow	communication through system bus	large size of input/output
Coprocessor	low	fast	limited size of coprocessor register-set	small size of input/output
Functional unit	low	very fast	limited size of processor registers	small size of input/output

*Comm' power: power consumption by data-storage (data buffer or
  or registers)
**Comm' speed: Communication speed between processor and RAA

addition, large data buffer of RAA can be used to support applications having large inputs/outputs. However, the speed improvement using the RAA may have to compensate for significant communication overhead between the processor and RAA through system bus as well as SRAM-based large data buffer in RAA consumes much power. Second type of organization involves the array connected with the processor as a 'Coprocessor' [12, 25, 61] shown in Figure 4.1(b). In this case, the standard processor does not change and the communication is faster than 'Attached IP' type interconnects because the co-processor register-set is used as data buffer of the RAA and the processor can access the register-set by coprocessor data transfer instructions. In addition, the register-set consumes less power than the data buffer of 'Attached IP'. Since the size of the register-set is fixed by the processor ISA, it creates performance bottleneck for registers-PE array traffic due to applications having large inputs/outputs run on the RAA. In the third type of organization, the array is placed inside the processor like a 'FU (Functional Unit)' [4, 7, 9, 24, 26] as shown in Figure 4.1(c). In this case, the instruction decoder issues special instructions to perform specific functions on the RAA as if it were one of the standard functional units of the processor. In this case, the communication speed is faster than 'Coprocessor' and power consumption of the data storage is less than 'Attached IP' because the processor register-set is used as data buffer of the RAA and the processor can directly access the register-set by the processor instructions. However, standard processor needs to be modified due to integration with RAA and its compiler should be also changed. The performance bottleneck is caused by limited size of the processor registers as in the case of 'Coprocessor' type organization. Table 10.1 shows a summary about advantage and disadvantage of three coupling types.

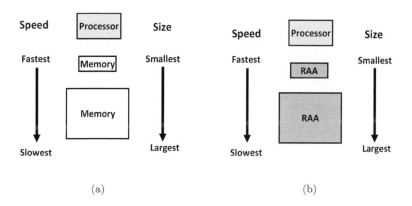

(a)   (b)

FIGURE 10.1: Analogy between memory and RAA-computing hierarchy: (a) memory; (b) RAA.

10.2.2   RAA-based Computing Hierarchy

As mentioned in the previous subsection, basic three types of RAA organizations show advantage and disadvantage according the input/output size of the applications. It shows the existing coupling structure with a conventional RAA cannot be flexible to support various applications with sacrificing performance. In addition, such an RAA structure cannot efficiently utilize PE arrays and data buffers leading to high power consumption. We hypothesize that if CGRA can maintain a computing hierarchy of its RAAs having different size and communication speed as shown in Figure 10.1(b), the CGRA-based embedded system can be optimized for its performance and power. It is because such a hierarchical arrangement of the RAA can optimize the communication latency and efficiently utilize functional resources of PE array in various applications. In this chapter, we propose a new CGRA-based architecture that supports such a RAA-based computing hierarchy.

10.3   Computing Hierarchy in CGRA

In order to implement efficient CGRA-based embedded systems, we propose a new computing hierarchy consisting of two computing blocks using two types of coupling structures together—'Attached IP' and 'Coprocessor'. In this organization, a general RAA having large size PE array is connected to a system bus and another is a small RAA composed of small PE array coupled with a processor through coprocessor interface. We call the small RAA

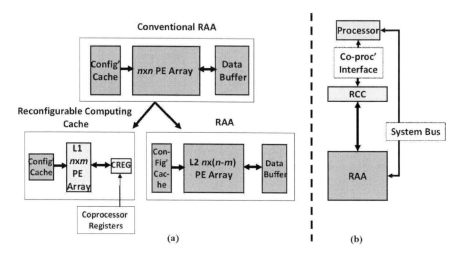

FIGURE 10.2: Computing hierarchy of CGRA: (a) size; (b) speed.

reconfigurable computing cache (RCC) because it plays important role in enhancing performance and power of the entire CGRA like data cache. The RCC and the RAA share critical resources and such a sharing structure provides efficient communication interface between two computing blocks. The proposed approach ensures that the RCC and the RAA are efficiently utilized to support variable size of inputs and outputs for variety of applications. In subsection 10.3.1 and 10.3.2, we describe computing hierarchy and resource sharing in RCC and RAA in detail. Then we show how to optimize computing flow based on reconfigurable computing cache according to the applications in subsection 10.3.3.

10.3.1   Computing Hierarchy—Size and Speed

A CGRA-based computing hierarchy is formed by splitting a conventional computing RAA block into two computing blocks—RCC with small PE array and RAA having large PE array as shown in Figure 10.2a. The RCC is coupled with coprocessor interface and the RAA is attached to a system bus as shown in Figure 10.2b. The RCC provides fast communication with the processor and offers low-power consumption by using coprocessor register-set and small array size. Therefore the RCC can enhance performance and reduce power consumption when small applications run on CGRA. If RCC is not sufficient to support computing requirements of applications, intermediate data from the RCC can be moved to the RAA through the interconnections as shown in Figure 10.3. Such interconnections between the two blocks offer flexibility in migrating computing demands from one to another. Such computing flow may help to optimize performance and power for the applications

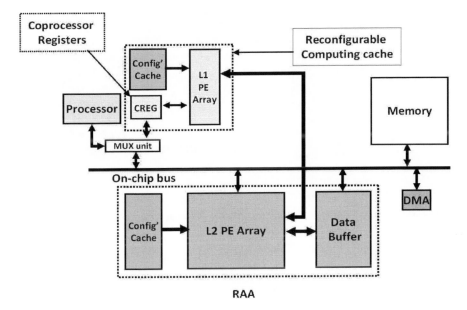

FIGURE 10.3: CGRA configuration with RCC and RAA.

having various sizes of inputs/outputs whereas the existing models show performance bottlenecks caused by the communication overheads or their limited sized data-storage as shown in Table 10.1. We have described the computing flow optimization in detail in subsection 10.3.3.

10.3.2   Resource Sharing in RCC and RAA

We have so far presented two factors (speed and size) in building computing hierarchy for CGRAs similar to memory hierarchy. It seems a small portion of RAA has been detached from large CGRA block and placed as the fast RCC block adjacent to the processor coupled with coprocessor interface. However, only considering two factors is not sufficient to design compact RCC for power and area benefits. This is because computing blocks can have diverse functionality which affects the system capabilities. The functionality of computing blocks is specified by functional resources of its PE such as adder, multiplier, shifter, logic operations etc. Therefore, it is necessary to examine how to select the functionalities of RCC and RAA. This leads to further studies on resource assignment/sharing between RCC and RAA. First of all, we can classify the functional resources into two groups: primitive resources and critical resources. Primitive resources are basic functional units such as adder/subtractor and logical operators. Critical resources are area/delay-critical ones such as multiplier and divider. Based on the classification, let us consider two cases of the functional resource configurations

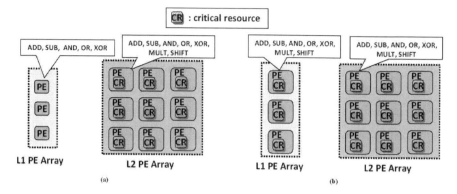

FIGURE 10.4: Two cases of functional resource assignment: (a) hierarchical functionality; (b) identical functionality.

as shown in Figure 10.4. Figure 10.4a shows hierarchical functionality that indicates L1 PE array has primitive resources and L2 PE array includes critical resources as well as primitive resources. The Figure 10.4b shows identical functionalities both in the L1 and L2 PE arrays. In the case of (a), the RCC with L1 PE array is relatively lightweight computing block compared to the RAA with L2 PE array. Therefore, the RCC can perform small applications having only primitive operations with low-power consumption. However, it causes 'lack of resource' problem when applications demand critical operations. In (b) L1 and L2 PE arrays have identical functionality with area and power overheads. To prevent such extreme cases, we propose resource sharing for the RCC and the RAA based on [39]. L1 and L2 PE array have the same primitive resources and shared the pipelined critical resources as shown in Figure 10.5. Here the RCC and the RAA basically perform the primitive operations and their functionality will include the critical operations using the shared resources. Figure 10.6 shows interconnection structure with shared critical resources along with RCC and RAA. PEs in the same row of the L1 and L2 array share the pipelined critical resources in the same manner as [39]. Such a structure avoids the 'lack of resource' problem in Figure 10.4a and this structure is more area and power-efficient than Figure 10.4a because the number of critical resources is reduced and the critical resources taken out of L1 and L2 PE array are not affected by unnecessary switching activity caused by other resources. In addition, interconnections for resource sharing can be also utilized for communication interface between the RCC and the RAA by adding multiplexer and de-multiplexer between front and end of the critical resources as shown in Figure 10.6b.

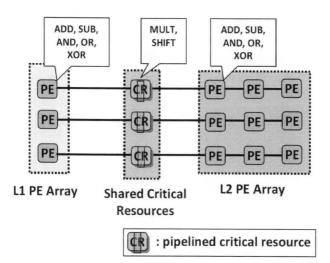

FIGURE 10.5: Critical resource sharing and pipelining in L1 and L2 PE array.

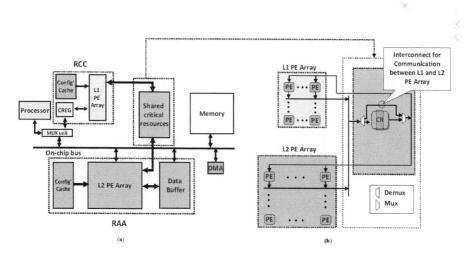

FIGURE 10.6: Interconnection structure among RCC, shared critical resources and L2 PE array: (a) entire structure; (b) interconnection structure.

10.3.3   Computing Flow Optimization

Based on the proposed CGRA structure, we can classify four cases of optimized computing flow to achieve low power and high performance. Figure 10.7 shows such four computing flows on the proposed CGRA according to variance of input and output size of applications—Subsection 10.4.1.1 shows that we can select the optimal case among the proposed computing flows for several applications with variance in their input/output size. All of the cases show that shared critical resources are used as needed because they are only utilized when applications have the operations requiring the critical resources. Figure 10.7(a) shows computing flow when application has the smallest inputs and outputs. In this case, only RCC functional units are used to execute the application while the RAA is disabled to reduce power consumption. However, if the application has larger inputs and outputs than (a), the computing flow can be extended to L2 PE array as shown in Figure 10.7(b). Even though L2 PE array is used for this case, data buffer of the RAA is not used because the coprocessor register-set (CREG) is sufficient to save all of the inputs or outputs. The next case is that when RAA is used with RCC because of large inputs and small outputs as shown in Figure 10.7(c). In this case, data buffer of the RAA receives inputs using DMA which is more efficient for overall performance than CREG. This is because insufficient CREG resource for large inputs causes performance bottleneck with heavy registers-PE array traffic. Therefore, the L2 PE array may be used first for running such application and the L1 PE array can be utilized for enhancing parallelized execution as needed. However, the outputs are stored on CREG because their size is small. Finally, Figure 10.7(d) shows a case of RAA used with L1 PE array with large inputs and outputs. To avoid heavy registers-PE array traffic by the large input/output size, the data buffer with DMA is used and L1 PE array can be optionally utilized for enhancing parallelized execution. In summary, the computing flow on the proposed CGRA can be adapted according to the input/output size of applications. It is more power-efficient than using a conventional CGRA by separated computing blocks with sharing critical resources. This way is only necessary computing blocks are utilized. In addition, computing flow with supporting two communication interfaces reduces power and enhances performance.

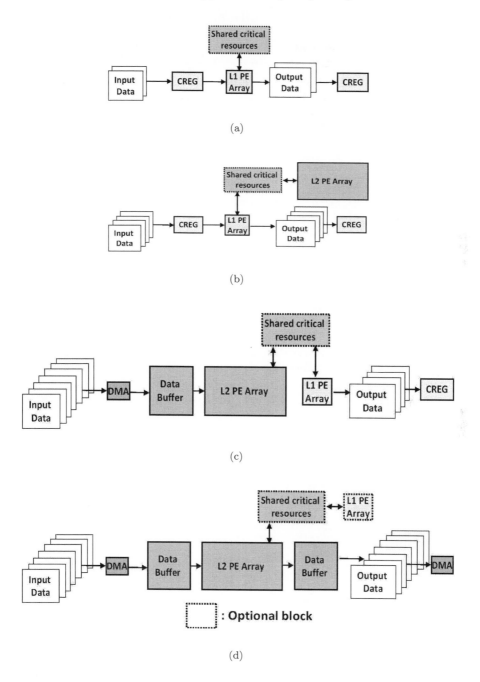

FIGURE 10.7: Four cases of computing flow according to the input/output size of application: (a) smallest inputs and outputs (STIO); (b) small inputs and outputs (SIO); (c) large inputs and small outputs (LISO); (d) large inputs and outputs (LIO).

TABLE 10.2:   Comparison of the architecture implementations

CGRA	PE array	Data storage
Attached IP	8x8 PE array	6KB data buffer
Coprocessor	8x8 PE array	256-byte coprocessor register-set
Proposed RCC-based	8x2 L1 PE array and 8x6 L2 PE array	4KB data butter and 256-byte coprocessor register-set
Leon2 processor [72] is used as main processor		

10.4   Experiments

10.4.1   Experimental Setup

10.4.1.1   Architecture Implementation

To demonstrate the effectiveness of the proposed RCC-based CGRA, we have designed three different organizations of CGRA with RT-level implementation using VHDL as shown in Table 10.2.

In addition, for resource sharing of RCC-based CGRA, two pipelined multipliers and two shifters are shared by PEs in the same row of L1 and L2 PE array whereas conventional two types of CGRA do not support such a resource sharing and pipelining. The architectures have been synthesized using Design Compiler [18] with 0.18 μm technology. PrimePower [18] has been used for gate-level simulation and power estimation. To obtain the power consumption data, we have used the applications in Table 10.2 for simulation with operation frequency of 100 MHz and typical case of 1.8 V Vdd and 27°C.

10.4.1.2   Evaluated Applications

Evaluated applications are composed of real multimedia applications and benchmarks. We have analyzed the input/output size and operation-types in the applications to identify specific computing flow in Figure 10.7. Table 10.3 shows the selected applications and the optimal computing flows for them.

10.4.2   Results

10.4.2.1   Area Cost Evaluation

Table 10.4 shows area cost evaluation for the two cases. 'Base 8x8' means 8x8 PE array included in 'Attached IP' and 'Coprocessor' type CGRA. 'Proposed' means L1 and L2 PE array included in the proposed RCC-based CGRA. Even though interconnection area of the proposed model increases because of resource sharing structure, entire area of the proposed one is reduced by 22.68% because it has less critical resources than base 8x8 PE array.

TABLE 10.3: Applications characteristics

Real Applications	SHR	Computing Flow	Benchmarks	SHR	Computing Flow
(H.263)8x8 DCT	✓	SIO	*256-point FFT	✓	LISO
(H.263)8x8 IDCT	✓	SIO	*256-tap FIR	✓	LISO
(H.263)8x8 QUANT	✓	SIO	*Complex Mult	✓	LISO
(H.263)8x8 DEQUANT	✓	SIO	**State	✓	STIO
(H.263)SAD	-	LISO	**Hydro	✓	STIO
(H.264)4x4 ITRANS	✓	STIO	**Tri-Diagonal	✓	LIO
(H.264)MSE	✓	LISO	**First-Diff	-	STIO
(H.264)MAE	-	LISO	**ICCG	✓	STIO
(H.264)16x16 DCT	✓	LISO	**Inner Product	✓	LIO
8x8*8x1Matrix-Vector Multiplication	✓	SIO	*: DSPstone benchmarks [34]		
16x16*16x1Matrix-Vector Multiplication	✓	LISO	**: Livermore loop benchmarks [6]		
8x8 Matrix Multiplication	✓	SIO	SHR: ✓ means critical resources are used for the application.		
16x16 Matrix Multiplication	✓	LISO	STIO: smallest inputs and outputs		
			SIO: small inputs and outputs		
			LISO: large inputs and small outputs		
			LIO: large inputs and outputs		

TABLE 10.4: Area cost comparison

PE Array	No' of PEs	No' of MULTs	No' of SHTs	Gate Equivalent			Reduction (%)
				Interconnect	Logic	Total	
Base 8x8	64	64	64	164908	494726	659635	-
Proposed	64	16	16	175434	334595	510029	22.68

10.4.2.2   Performance Evaluation

The synthesis results show that the proposed PE array has reduced critical path delay (5.12 ns) compared to the base PE array (8.96 ns). This is because pipelined multipliers are excluded from the original set of critical paths. Based on the synthesis results, we evaluate execution times of the selected applications on three cases of CGRA as shown in Figure 10.8. The execution times include communication time between memory/processor and the RAA or RCC. Each application is executed on the RCC-based CGRA in the manner of selected computing flow as shown in Table 10.3—all of the applications are classified under 4 cases of computing flow (STIO, SIO, LISO and LIO). In the case of STIO and SIO, performance improvement compared with 'Coprocesssor' type is relatively less (30.31%~37.03%) than LIO and LISO (60.94%~72.92%). This is because the improvements of STIO and SIO are achieved by only reduced critical path delay whereas the improvements of LIO or LISO are achieved by avoiding heavy coprocessor registers-PE array traffic as well as reduced critical path delay. However, compared with 'Attached-IP' type, STIO and SIO achieve much more performance improvement (56.60%~67.90%) whereas LISO and LIO show the improvement of (42.05%~59.85%). This is because STIO and SIO do not use data buffer of the RAA causing communication overhead on system bus.

10.4.2.3   Power Evaluation

Figure 10.9 shows the comparison of power consumptions in three different organizations of CGRA. First of all, the proposed L1 and L2 PE array is more power-efficient than the base PE array because of the reduced critical resources. With such a power-efficient PE array, the amount of power saving depends on the selected computing flow for the application. The most power-efficient computing flow is STIO that shows relatively much power saving (40.44%~55.55%) compared to other cases (7.93%~29.67%) because the STIO does not use the RAA—especially, 'First_Diff' shows the highest power saving ratio of 51.71%/55.55% because of not using the shared critical resources. The next power-efficient model is SIO showing power saving (23.67%~29.67%). This is because the SIO computing flow does not use data buffer of the RAA whereas LISO (7.93%~26.03%) and LIO (17.13%~22.91%) utilizes the data buffer for input data or output data. Finally, power saving of LISO and LIO is mostly achieved by reduced critical resources and by not activating L1 PE array.

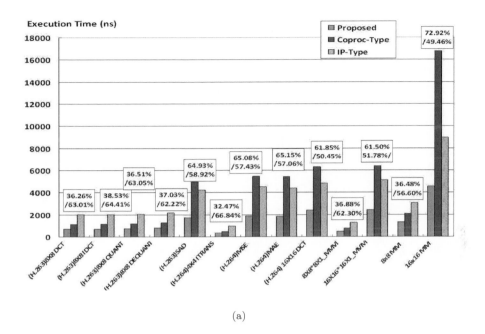

(a)

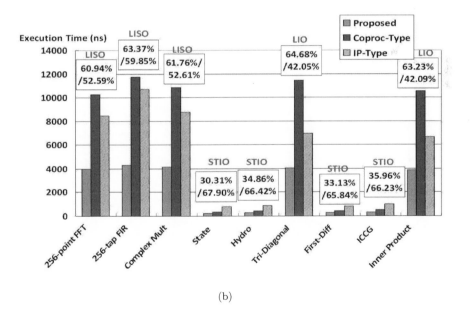

(b)

FIGURE 10.8: Performance comparison: (a) real applications; (b) benchmarks (A%/B%: A% means reduced execution time ratio compared with Coproc-Type and B% means reduced execution time ratio compared with IP-Type).

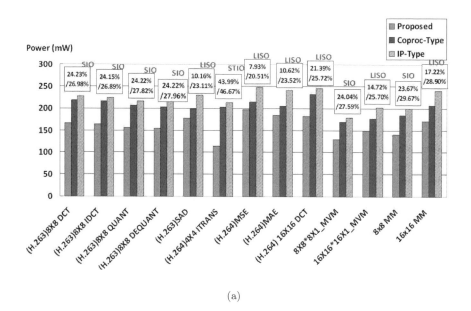

(a)

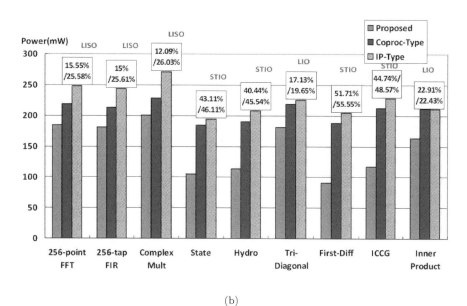

(b)

FIGURE 10.9: Power comparison: (a) real applications; (b) benchmarks (A%/B%: A% means power saving ratio compared with Coproc-Type, and B% means power saving ratio compared with IP-Type).

10.5   Summary

Coarse-grained reconfigurable architectures have emerged as a suitable solution for embedded systems because it aims to achieve high performance and flexibility. However, the CGRA has been considered as prohibitive one due to its significant area/power overhead and performance bottleneck. This is because existing CGRAs cannot efficiently utilize large reconfigurable computing block of many processing elements and data buffer wasting area and power. In addition, fixed communication structure between the computing block and processor cannot guarantee good performance for various applications. To overcome the limitations, in this chapter, we introduce a new computing hierarchy consisting of two reconfigurable computing blocks with two types of communication structure together. In addition, the two computing blocks have shared critical resources. Such a sharing structure provides efficient communication interface between them with reducing overall area. Based on the proposed architecture, optimized computing flows have been implemented according to the varying applications for low power and high performance. Experiments with implementation of several applications on the new hierarchical CGRA demonstrate the effectiveness of our proposed approach. The proposed approach reduces area by up to 22.68%, execution time by up to 72.92% and power by up to 55.55% when compared with the existing undivided arrays in CGRA architectures.

Chapter 11

Integrated Approach to Optimize CGRA

11.1 Combination among the Cost-Effective CGRA Design Schemes

From Chapter 6 to Chapter 10, we have proposed the cost-effective CGRA design schemes and such schemes can be combined with each other to optimize CGRA in terms of area, power and performance. Figure 11.1 shows combination flow of the proposed design schemes. The flow shows possible scheme combinations for CGRA design. Each arrow of the flow shows a possible integration between two design schemes. The possible scheme combinations can be found by tracing in the arrow directions. The combination flow can be classified into two cases according to the computation model of CGRA. In the case of temporal mapping, *low-power reconfiguration technique by reusable context pipelining* (Chapter 6) can be selected whereas *cost-effective array fabric* (Chapter 9) is applicable to the spatial mapping. This is because two design schemes have been devised while keeping the characteristics of spatial mapping and temporal mapping—we spatially spread the operations in the data flows over the array space in the design scheme of the *cost-effective array fabric* whereas *reusable context pipelining* spread the operations over time for each column to implement temporal loop pipelining. Therefore, even though two design schemes cannot be merged, any combination of a design scheme in Chapter 6 or Chapter 9 with the remaining three schemes is possible.

11.2 Case Study for Integrated Approach

11.2.1 An CGRA Design Example Merging Three Design Schemes

To demonstrate the effectiveness of the integrated approach, we have designed a RAA combining three design schemes as shown in Figure 11.2 with

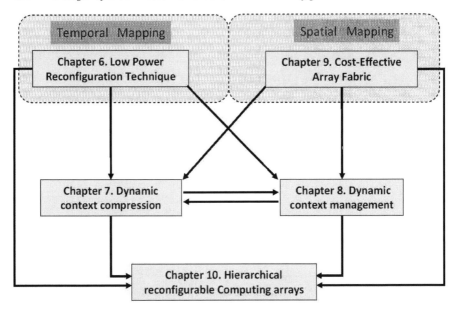

FIGURE 11.1: Combination flow of the proposed design schemes.

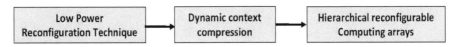

FIGURE 11.2: A combination example combining three design schemes.

RT-level implementation using VHDL. The architectures have been synthesized using Design Compiler [18] with 0.18 μm technology. PrimePower [18] has been used for gate-level simulation and power estimation. To obtain the power consumption data, we have used the same applications shown in the previous chapters for simulation with operation frequency of 100 MHz and typical case of 1.8 V Vdd and 27°C.

11.2.2   Results

11.2.2.1   Area and Performance Evaluation

Table. 11.1 shows area cost evaluation of each component for the base RAA as specified in Chapter 4 and the integrated RAA combining three design schemes. In the case of configuration cache, area is reduced by 16.79%—even though *dynamic context compression* increases area as shown in Chapter 8, *low-power reconfiguration technique* offsets the increased area with reduced size of the configuration cache. Area of the PE array and frame buffer are also reduced by 17.27% and 30% because *hierarchical reconfigurable comput-*

TABLE 11.1: Area reduction ratio by integrated RAA

Component	Gate Equivalent		Reduction (%)
	Base	Integrated	
Configuration Cache	150012	124824	16.79
PE Array	659635	510029	22.68
Frame Buffer	129086	90329	30.00
Entire RAA	942742	760869	**23.07**

ing arrays supports critical resource sharing with the reduced size of the frame buffer. Therefore, the area reduction ratio of the entire RAA is 23.07% compared to the base RAA. The synthesis results show that the integrated RAA has reduced critical path delay (5.12 ns) compared to the base RAA (8.96 ns). This is because *dynamic context management* and *low-power reconfiguration technique* don't affect the original critical path delay and pipelined multipliers are excluded from the original set of critical paths by *hierarchical reconfigurable computing arrays*. In addition, execution time evaluation of the applications shows the same results in Chapter 10—performance enhancement of 42.05%~67.90% compared with the IP-type base RAA.

11.2.2.2 Power Evaluation

To verify the synergy effect of the integrated approach, we have evaluated power consumption for the five cases:

- *a*. Base RAA

- *b*. RAA with *low-power reconfiguration technique*

- *c*. RAA with *dynamic context compression*

- *d*. RAA with *hierarchical reconfigurable computing array*

- *e*. Integrated RAA.

Table 11.2 shows entire power comparison among the five cases. Each design scheme (*b*, *c* and *d*) does not reduce much power of entire RAA— 26.54%~47.6% in *b*, 13.77%~21.48% in *c* and 11.09%~30.19% in *d*. However, the integrated RAA save much power (44.65%~71.29%) because each component of the RAA is optimized by the individual design scheme.

11.3   Potential Combinations and Expected Outcomes

As mentioned in Section 11.1, any combination of a design scheme limited by the computation model with the remaining four schemes is possible and we

TABLE 11.2: Entire power comparison

kernels	[a]Base	[b]Low Power Reconfig'		[c]Context compression		[d]Hierarchical Array		[e]Integrated	
	[f]P	[f]P	[g]R	[f]P	[g]R	[f]P	[g]R	[f]P	[g]R
[h]First_Diff	376.17	232.48	38.2	309.37	17.76	262.62	30.19	108.01	71.29
[h]Tri-Diagonal	400.19	257.59	35.63	331.01	17.29	355.79	11.09	200.65	49.86
[h]State	356.08	228.45	35.84	294.23	17.37	266.23	25.23	125.71	64.7
[h]Hydro	356.47	240.64	32.49	299.74	15.91	261.64	26.6	133.41	62.57
[h]ICCG	434.45	261.29	39.86	354.33	18.44	323.39	25.56	137.52	68.35
[h]Inner Product	328.54	240.57	26.78	283.3	13.77	281.27	14.39	181.83	44.65
[i]24-Taps FIR	471.44	274.99	41.67	383.44	18.67	408.98	13.25	200.5	57.47
[j]MVM	405.7	212.58	47.6	318.56	21.48	356.56	12.11	150.25	62.97
Mult in FFT	423.59	287.67	32.09	355.19	16.15	360.12	14.98	208.78	50.71
[k]Comlex Mult	452	304.19	32.7	381.55	15.59	381.38	15.62	220.77	51.16
[l]TRANS	417.95	283.06	32.27	338.37	19.04	318.49	23.8	156.42	62.57
[m]DCT	417.33	264.89	36.53	347.17	16.81	356	14.7	189.68	54.55
[m]IDCT	412.91	263.45	36.2	343.42	16.83	352.55	14.62	188.71	54.3
[m]SAD	415.27	305.05	26.54	343.04	17.39	362.12	12.8	222.63	46.39
[m]Quant	401.35	255.77	36.27	333.63	16.87	341.22	14.98	181.14	54.87
[m]Dequant	401.64	252.3	37.18	332.63	17.18	341.85	14.89	178.38	55.59

[a]Base RAA (configuration cache+frame buffer+PE array),
[b]RAA with low-power reconfiguration technique, [c]RAA with dynamic context compression,
[d]RAA with hierarchical reconfigurable computing array,
[e]RAA combining three scheme, [f]Power consumption of RAA (mW),
[g]Power reduction ratio of entire RAA compared with BASE (%),
[h]Livermore loop benchmark suite, [i]DSPstone benchmark suite,
[j]Matrix-vector multiplication, [k]Kernels in AAC,
[l]Kernels in H.264, [m]Kernels in H.263

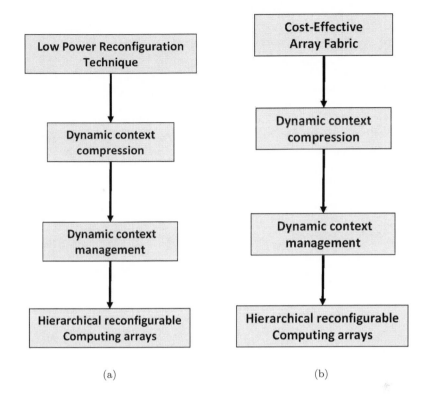

(a)   (b)

FIGURE 11.3: Potential combination of multiple design schemes: (a) power optimization for the configuration cache; (b) area/power optimization of the PE array.

can consider two cases of the maximum combinations—one is the maximum power optimization for the configuration cache and another is area/power optimization of the PE array. Figure 11.3 shows such two cases of combinations. In the case of Figure 11.3(a), all of the design schemes reducing power in configuration cache are merged with *hierarchical reconfigurable computing arrays*. Therefore, power saving of the configuration cache can be optimized based on the computation model of the temporal mapping. The second case is area/power optimization of the PE array as shown in Figure 11.3(b). Compared with (a), instead of *low-power reconfiguration technique*, the design scheme of *cost-effective array fabric* is combined with other design schemes. In this case, the area/power of the PE array can be optimized by reducing the number of PEs (*cost-effective array fabric*) and sharing critical-resource (*hierarchical reconfigurable computing arrays*).

11.4   Summary

In this chapter, we present integrated approach to merge the multiple design schemes presented in the previous chapters. A case study is shown to verify the synergy effect of combining the multiple design schemes. Experimental results show that the integrated approach reduces area by 23.07% of entire RAA and power by up to 72% when compared with the conventional RAA. In addition, we discuss potential combinations among the proposed design schemes and their expected outcomes.

Bibliography

[1] M. Ahn, J. Yoon, Y. Paek, Y. Kim, M. Kiemb, and K. Choi. A spatial mapping algorithm for heterogeneous coarse-grained reconfigurable architectures. In *Proceedings of Design Automation and Test in Europe Conference*, pages 262–268, 2006.

[2] T. Arai, I. Kuroda, K. Nadehara, and K. Suzuki. Automated target recognition on SPLASH 2. In *Proceedings of IEEE Symposium on Field-Programmable Custom Computing Machines*, pages 192–200, 1997.

[3] T. Arai, I. Kuroda, K. Nadehara, and K. Suzuki. V830r /av: Embedded multimedia superscalar RISC processor. In *Proceedings of IEEE Micro*, pages 36–47, 1998.

[4] C. Arbelo, A. Kanstein, S. Lopez, J.F. Lopez, M. Berekovic, R. Sarmiento, and J.-Y. Mignolet. Mapping control-intensive video kernels onto a coarse-grain re-configurable architecture: the h.264/avc deblocking filter. In *Proceedings of Design Automation and Test in Europe Conference*, pages 642–649, 2007.

[5] TMS320C6000 Assembly Benchmarks at Texas Instruments. Website: http://www.ti.com/sc/docs/products/dsp/c6000/benchmarks.

[6] Netlib Repository at the Oak Ridge National Laboratory. Website: http://www.netlib.org/benchmark/livermorec.

[7] N. Bansal, S. Gupta, N. Dutt, and A. Nicolau. Analysis of the performance of coarse-grain reconfigurable architectures with different processing element configurations. In *Proceedings of Workshop on Application Specific Processors*, 2003.

[8] N. Bansal, S. Gupta, N. Dutt, A. Nicolau, and R. Gupta. Interconnect-aware mapping of applications to coarse-grain reconfigurable architectures. In *Proceedings of International Conference on Field Programmable Logic and Applications*, pages 891–899, 2004.

[9] F. Barat, M. Jayapala, T. Vander A. Corporaal, G. Deconinck, and R. Lauwereins. Low power coarse-grained reconfigurable instruction set processor. In *Proceedings of International Conference on Field Programmable Logic and Applications*, pages 230–239, 2003.

[10] F. Barat and R. Lauwereins. Reconfigurable instruction set processors: A survey. In *Proceedings of International Workshop on Rapid System Prototyping*, pages 168–173, 2000.

[11] J. Becker and M. Vorbach. Architecture, memory and interface technology integration of an industrial/academic configurable system-on-chip (CSoC). In *Proceedings of IEEE Computer Society Annual Symposium on VLSI*, pages 107–112, 2003.

[12] T. Callahan, J. Hauser, and J. Wawrzynek. The Garp architecture and C compiler. *IEEE Computer*, 33(4):62–69, 2000.

[13] K. Chia, H.J. Kim, S. Lansing, W.H. Mangione-Smith, and J. Villasensor. High-performance automatic target recognition through data-specific VLSI. *IEEE Transactions on Very Large Scale Integration Systems*, 6(3):364–371, 1998.

[14] Y. Chou, P. Pillai, H. Schmit, and J. Shen. PipeRench implementation of the instruction path coprocessor. In *Proceedings of Annual IEEE/ACM International Symposium on Microarchitecture*, pages 147–158, 2000.

[15] J. Cocke. Global common sub expression elimination. In *Proceedings of Symposium on Compiler Construction, ACM SIGPLAN Notices*, pages 850–856, 1970.

[16] ARM Corporation. Website: http://www.arm.com/arm/amba.

[17] Model Technology Corporation. Website: http://www.model.com.

[18] Synopsys Corporation. Website: http://www.synopsys.com.

[19] A. Deledd, C. Mucci, A. Vitkovski, M. Kuehnle, F. Ries, M. Huebner, J. Becker, P. Bonnot, A. Grasset, P. Millet, M. Coppola, L. Pieralisi, R. Locatelli, and G. Maruccia. Design of a hw/sw communication infrastructure for a heterogeneous reconfigurable processor. In *Proceedings of Design, Automation, and Test in Europe Conference*, pages 1352–1357, 2008.

[20] G. Dimitroulakos, N. Kostaras, M. Galanis, and C. Goutis. Compiler assisted architectural exploration for coarse grained reconfigurable arrays. In *Proceedings of Great Lakes Symposium on VLSI*, pages 164–167, 2007.

[21] A. Kanstein, F. Bouwens, M. Berekovic, and G. Gaydadjiev. Architectural exploration of the ADRES coarse-grained reconfigurable array. In *Proceedings of International Workshop on Applied Reconfigurable Computing*, pages 1–13, 2007.

[22] Application Notes for Pentium MMX. Website: http://developer.intel.com/drg/mmx/appnotes.

[23] P. Heysters, G. Rauwerda, and G. Smit. Towards software defined radios using coarse-grained reconfigurable hardware. *IEEE Transactions on Very Large Scale Integration Systems*, 16(1):3–13, 2008.

[24] M. Galanis, G. Dimitroulakos, and C. Goutis. Speedups and energy savings of microprocessor platforms with a coarse-grained reconfigurable datapath. In *Proceedings of International Parallel and Distributed Processing Symposium*, pages 1–8, 2007.

[25] M. Galanis, G. Dimitroulakos, S. Tragoudas, and C. Goutis. Speedups in embedded systems with a high-performance coprocessor datapath. *ACM Transactions on Design Automation of Electronic Systems*, 12(35):1–22, 2007.

[26] M. Galanis and C. Goutis. Speedups from extending embedded processors with a high-performance coarse-grained reconfigurable data-path. *Journal of Systems Architecture—Embedded Systems Design*, 49(2):479–490, 2008.

[27] S.C. Goldstein, H. Schmit, M. Budiu, S. Cadambi, M. Moe, and R.R Taylor. PipeRench: a reconfigurable architecture and compiler. *IEEE Computer*, 33(4):70–77, 2000.

[28] F. Hannig, H. Dutta, and J. Teich. Mapping of regular nested loop programs to coarse-grained reconfigurable arrays—constraints and methodology. In *Proceedings of IEEE International Parallel and Distributed Processing Symp*, pages 148–155, 2004.

[29] F. Hanning, H. Dutta, and J. Teich. Regular mapping for coarse-grained reconfigurable architectures. In *Proceedings of IEEE International Conference on Acoustics, Speech, and Signal Processing*, pages 57–60, 2004.

[30] R. Hartenstein. A decade of reconfigurable computing: a visionary retrospective. In *Proceedings of Design Automation and Test in Europe Confererence*, pages 642–649, 2001.

[31] R. Hartenstein, M. Herz, T. Hoffmann, and U. Nageldinger. KressArray Xplorer: a new CAD environment to optimize reconfigurable datapath array architectures. In *Proceedings of Asia and South Pacific Design Automation Conference*, pages 163–168, 2000.

[32] M. Hartmann, V. Pantazis, T. Vander Aa, M. Berekovic, C. Hochberger, and B. de Sutter. Still image processing on coarse-grained reconfigurable array architectures. In *Proceedings of IEEE/ACM/IFIP Workshop on Embedded Systems for Real-Time Multimedia*, pages 67–72, 2007.

[33] C. Hsieh and T. Lin. VLSI architecture for block-matching motion estimation algorithm. *IEEE Transactions on Circuits and Systems for Video Technology*, 2(2):169–175, 1992.

[34] Germany Institute for Integrated Signal Processing Systems, Aachen. Website: http://www.ert.rwth-aachen.de/projekte/tools/dspstone.

[35] M. Jo, V. Arava, H. Yang, and K. Choi. Implementation of floating-point operations for 3d graphics on a coarse-grained reconfigurable architecture. In *Proceedings of IEEE International SoC Conference*, pages 127–130, 2007.

[36] S. Keutzer, S. Tjiang, and S. Devadas. A new viewpoint on code generation for directed acyclic graphs. *ACM Transactions on Design Automation of Electronic Systems*, 3(1):51–75, 1998.

[37] S. Khawam, T. Arslan, and F. Westall. Synthesizable reconfigurable array targeting distributed arithmetic for system-on-chip applications. In *Proceedings of IEEE International Parallel and Distributed Processing Symposium*, pages 150–157, 2004.

[38] Y. Kim, M. Kiemb, and K. Choi. Efficient design space exploration for domain-specific optimization of coarse-grained reconfigurable architecture. In *Proceedings of IEEE IEEK SoC Design Conference*, pages 19–24, 2005.

[39] Y. Kim, M. Kiemb, C. Park, J. Jung, and K. Choi. Resource sharing and pipelining in coarse-grained reconfigurable architecture for domain-specific optimization. In *Proceedings of Design Automation and Test in Europe Conference*, pages 12–17, 2005.

[40] Y. Kim, J. Lee, J. Junng, S. Kang, and K. Choi. Design of coarse-grained reconfigurable hardware. In *Proceedings of IEEK SoC Design Conference*, pages 312–317, 2004.

[41] Y. Kim and R. Mahapatra. Dynamically compressible context architecture for low power coarse-grained reconfigurable array. In *Proceedings of Int. Conference on Computer Design*, pages 295–400, 2007.

[42] Y. Kim and R. Mahapatra. A new array fabric for coarse-grained reconfigurable architecture. In *Proceedings of EuroMicro Conference on Digital System Design*, pages 584–591, 2008.

[43] Y. Kim and R. Mahapatra. Reusable context pipelining for low power coarse-grained reconfigurable architecture. In *Proceedings of International Parallel and Distributed Processing Symposium*, pages 1–8, 2008.

[44] Y. Kim and R. Mahapatra. Dynamic context management for low power coarse-grained reconfigurable architecture. In *Proceedings of ACM Great Lake Symposium on VLSI*, pages 33–38, 2009.

[45] Y. Kim and R. Mahapatra. Hierarchical reconfigurable computing arrays for efficient CGRA-based embedded systems. In *Proceedings of Design Automation Conference*, pages 1–6, 2009.

[46] Y. Kim, C. Park, S. Kang, H. Song, J. Jung, and K. Choi. Design and evaluation of coarse-grained reconfigurable architecture. In *Proceedings of International SoC Design Conference*, pages 227–230, 2004.

[47] Y. Kim, I. Park, K. Choi, and Y. Paek. Power-conscious configuration cache structure and code mapping for coarse-grained reconfigurable architecture. In *Proceedings of International Symposium on Low Power Electronics and Design*, pages 310–315, 2006.

[48] A. Lambrechts, P. Raghavan, and M. Jayapala. Energy-aware interconnect-exploration of coarse-grained reconfigurable processors. In *Proceedings of Workshop on Application Specific Processors*, pages 150–157, 2005.

[49] M. Lanuzza, M. Margala, and P. Corsonello. Cost-effective low-power processor-in-memory-based reconfigurable datapath for multimedia applications. In *Proceedings of International Symposium on Low Power Electronics and Design*, pages 161–166, 2005.

[50] G. Lee, S. Lee, and K. Choi. Automatic mapping of application to coarse-grained reconfigurable architecture based on high-level synthesis techniques. In *Proceedings of IEEE International SoC Design Conference*, pages 395–398, 2008.

[51] J. Lee, K. Choi, and N. Dutt. Mapping loops on coarse-grained reconfigurable architectures using memory operation sharing. *Technical Report, Center for Embedded Computer Systems (CECS), University of California, Irvine*, pages 2–34, 2002.

[52] J. Lee, K. Choi, and N. Dutt. An algorithm for mapping loops onto coarse-grained reconfigurable architectures. In *Proceedings of ACM Workshop on Languages, Compilers, Tools for Embedded Systems*, pages 183–188, 2003.

[53] J. Lee, K. Choi, and N. Dutt. An algorithm for mapping loops onto coarse-grained reconfigurable architectures. *ACM Sigplan Notices*, 38(7):183–188, 2003.

[54] J. Lee, K. Choi, and N. Dutt. Design space exploration of reconfigurable ALU array (RAA) architectures. In *Proceedings of IEEE SOC Design Conference*, pages 302–307, 2003.

[55] J. Lee, K. Choi, and N. Dutt. Evaluating memory architectures for media applications on coarse-grained reconfigurable architectures. In *Proceedings of IEEE International Conference on Application-Specific Systems, Architectures, and Processors*, pages 166–176, 2003.

[56] J. Lee, K. Choi, and N. Dutt. Evaluating memory architectures for media applications on coarse-grained reconfigurable architectures. *International Journal of Embedded Systems*, 3(3):119–127, 2008.

[57] W. Lee and J. Kim. H.264 implementation with embedded reconfigurable architecture. In *Proceedings of the Sixth IEEE International Conference on Computer and Information Technology, 2006*, pages 247–251, 2006.

[58] G. Lu. Modeling, Implementation and Scalability of the MorphoSys Dynamically Reconfigurable Computing Architecture. *Ph.D. Dissertation, Electrical and Computer Engineering, University of California, Irvine*, pages 1–221, 2000.

[59] B. Mei, F.J. Veredas, and B. Masschelein. Mapping an h.264/avc decoder onto the ADRES reconfigurable architecture. In *Proceedings of International Conference on Field Programmable Logic and Applications*, pages 622–625, 2005.

[60] B. Mei, S. Vernalde, D. Verkest, and R. Lauwereins. Design methodology for a tightly coupled VLIW/reconfigurable matrix architecture: a case study. In *Proceedings of Design Automation and Test in Europe Conference*, pages 1224–1229, 2004.

[61] T. Miyamori and K. Olukotun. A quantitative analysis of reconfigurable coprocessors for multimedia applications. In *Proceedings of IEEE Symposium on FPGAs for Custom Computing Machines*, pages 15–17, 1998.

[62] M. Myjak and J. Delgado-Frias. A medium-grain reconfigurable architecture for DSP: VLSI design, benchmark mapping, and performance. *IEEE Transactions on Very Large Scale Integration Systems*, 16(1):14–23, 2008.

[63] N. Dutt, A. Nicolau, N. Bansal, S. Gupta, and R. Gupta. Network topology exploration of mesh-based coarse-grain reconfigurable architectures. In *Proceedings of Design Automation and Test in Europe Conference*, pages 474–479, 2004.

[64] S.H. Nam, J.S. Beak, T.Y. Lee, and M.K. Lee. A VLSI designs for motion compensation block matching algorithm. In *Proceedings of IEEE ASIC Conference*, pages 254–257, 1994.

[65] C. Park, Y. Kim, and K. Choi. Domain-specific optimization of reconfigurable array architecture. In *Proceedings of US-Korea Conference on Science, Technology, & Entrepreneurship*, 2005.

[66] H. Park, K. Fan, M. Kudlur, and S. Mahlke. Modulo graph embedding: Mapping applications onto coarse-grained reconfigurable architectures. In *Proceedings of International Conference on Compilers, Architecture, and Synthesis for Embedded Systems*, pages 136–146, 2006.

[67] I. Park, Y. Kim, M. Jo, and K. Choi. Chip implementation of power conscious configuration cache for coarse-grained reconfigurable architecture. In *Proceedings of the 15th Korean Conference on Semiconductors*, pages 527–528, 2008.

[68] I. Park, Y. Kim, C. Park, J. Son, M. Jo, and K. Choi. Chip implementation of a coarse-grained reconfigurable architecture. In *Proceedings of IEEE International SoC Design Conference*, pages 628–629, 2006.

[69] A. Poon. An energy-efficient reconfigurable baseband processor for wireless communications. *IEEE Transactions on Very Large Scale Integration Systems*, 15(3):319–327, 2007.

[70] A. Poon. An energy-efficient reconfigurable baseband processor for wireless communications. *IEEE Transactions on Very Large Scale Integration Systems*, 15(3):319–327, 2007.

[71] R. Rau. *Iterative modulo scheduling*. Technical Report, Hewlett-Packard Lab: HPL-94-115, 1995.

[72] Gaisler Research. Website: http://www.gaisler.com/cms.

[73] S. Salomao, V. Alves, and E.C. Filho. A high performance VLSI cryptographic chip. In *Proceedings of IEEE ASIC Conference*, pages 7–13, 1998.

[74] H. Schmit, D. Whelihan, M. Moe, A. Tsai, B. Levine, and R. Taylor. PipeRench: A virtualized programmable datapath in 0.18 micron technology. In *Proceedings of IEEE Custom Integrated Circuits Conference*, pages 63–66, 2002.

[75] H. Singh, M. Lee, G. Lu, F. Kurdahi, N. Bagherzadeh, and E. Filho. MorphoSys: An integrated reconfigurable system for data-parallel and computation-intensive applications. *IEEE Transactions on Computers*, 49(5):465–481, 2000.

[76] N. Suzuki, S. Kurotaki, M. Suzuki, N. Kaneko, Y. Yamada, K. Deguchi, Y. Hasegawa, H. Amano, K. Anjo, M. Motomura, K. Wakabayashi, T. Toi, and T. Awashima. Implementing and evaluating stream applications on the dynamically reconfigurable processor. In *Proceedings of Field-Programmable Custom Computing Machines*, pages 328–329, 2004.

[77] G. Venkataramani, W. Najjar, F. Kurdahi, N. Bagherzadeh, and W. Bohm. A compiler framework for mapping applications to a coarse-grained reconfigurable computer architecture. In *Proceedings of International Conference on Compilers, Architecture, and Synthesis for Embedded Systems*, pages 116–125, 2001.

[78] F. Vererdas, M. Scheppler, W. Moffat, and B. Mei. Custom implementation of the coarse-grained reconfigurable ADRES architecture for multimedia purposes. In *Proceedings of International Conference on Field Programmable Logic and Applications*, pages 106–111, 2005.

[79] K.M. Yang, M.T. Sun, and L. Wu. A family of VLSI designs for motion compensation block matching algorithm. *IEEE Transactions on Circuits and Systems*, 36(10):1,317–1,325, 1989.

[80] J. Yoon, Y. Kim, M. Ahn, Y. Paek, and K. Choi. Temporal mapping for loop pipelining on a MIMD style coarse-grained reconfigurable architecture. In *Proceedings of IEEE International SoC Design Conference*, pages 1–8, 2006.

[81] J. Yoon, A. Shrivastava, S. Park, M. Ahn, R. Jeyapaul, and Y. Paek. SPKM : A novel graph drawing based algorithm for application mapping onto coarse-grained reconfigurable architectures. In *Proceedings of Asia and South Pacific Design Automation Conference*, pages 776–782, 2008.

[82] H. Zhang, M. Wan, V. George, and J. Rabaey. Interconnect architecture exploration for low-energy reconfigurable single-chip DSPS. In *Proceedings of VLSI' 99*, pages 150–157, 1999.

Index

Note: page numbers in italics indicate figures; page numbers in bold indicate tables.

A

ADRES (architecture for dynamically reconfigurable embedded systems), 20–22, 39–40, 51–52, 72–73
  architecture, 20–22, *54*, *55*
  breakdown of power consumption in CGRA and, 72–73
  core, *21*
  H.264/AVC Decoder, 51
  image processing algorithm, 51–52
  performance evaluation and, 51–52
  processor power consumption with IDCT, *75*
ADRES - DRESC, 29
  design space exploration and, 29
ADRES - Modolo Scheduling, 35–36
ALU-dependent fields, control signals and, 107
Amdahl's Law, 74
application mapping flow, 91–93
  for base architecture, *93*
  context rearrangement, 94–95
  for proposed architecture, 92–93
  temporal mapping algorithm, 92–93
application speedups, execution times and, *60*
applications
  characteristics of, *58*, **164**, **177**
  computation-intensive, 143

data parallel, 143
  performance evaluation and, **164**
applications' characteristics performance evaluation, **164**
application-specific integrated circuit (ASIC), 17
architecture
  ADRES, 20–22
  MorphoSys, 8–13
  PACT-XPP, 17–19
  PipeRench, 25–26
  RaPiD, 23–24
  REMARC, 14–16
architecture implementations, 176
  comparison of, **176**
architecture specification, of base and proposed architecture, **97**
area cost
  breakdown for CGRA, *68*
  comparison, **177**
  evaluation, 176–177
area overhead
  by dynamic context compression, **120**
  by dynamic context management, **133**
  dynamic context management and, **133**
area reduction ratio
  by integrated RAA, **185**
  by RSPA and NAF, **163**
automatic target recognition, 48

197

Printed and bound by CPI Group (UK) Ltd, Croydon, CR0 4YY

25/10/2024

01779224-0002